I0482775

NISTIR 7556

Report of Findings: Use of Language in Ballot Instructions

Janice (Ginny) Redish, Ph.D.
Redish & Associates

Dana E. Chisnell
UsabilityWorks

Ethan Newby
Newby Research

Sharon J. Laskowski
Information Access Division
Information Technology Laboratory
National Institute of Standards and Technology

Svetlana Z. Lowry
Information Access Division
Information Technology Laboratory
National Institute of Standards and Technology

December 2008

U.S. Department of Commerce
Carlos M. Gutierrez, Secretary
National Institute of Standards and Technology
James M. Turner, Deputy Director

NISTIR 7556

Report of Findings:
Use of Language in Ballot Instructions

Janice (Ginny) Redish, Ph.D.
Redish & Associates

Dana Chisnell
UsabilityWorks

Ethan Newby
Newby Research

Sharon J. Laskowski
Information Access Division
Information Technology Laboratory
National Institute of Standards and Technology

Svetlana Z. Lowry
Information Access Division
Information Technology Laboratory
National Institute of Standards and Technology

December 2008

This document has been prepared by the National Institute of Standards and Technology (NIST) and describes research in support of test methods and materials for the Election Assistance Commission's next iteration of the Voluntary Voting System Guidelines (VVSG). It does not represent a consensus view or recommendation from NIST, nor does it represent any policy positions of NIST.

Certain commercial entities, equipment, or material may be identified in the document in order to describe an experimental procedure or concept adequately. Such identification is not intended to imply recommendation or endorsement by the National Institute of Standards and Technology, nor is it intended to imply that these entities, materials, or equipment are necessarily the best available for the purpose.

Table of Contents

Report of Findings:
Use of Language in Ballot Instructions

Highlights and Summary

In a study of 45 voters in three geographic locations comparing a ballot with traditional language instructions to a ballot with plain language instructions, we collected both performance and preference data. The traditional language was language commonly found in actual ballots across the United States. A detailed explanation of plain language can be found in Part 1 of the full report.

For performance data, participants voted on ballots that differed only in the wording and placement of instructions: Ballot A, traditional instructions; Ballot B, plain language instructions. Half of the participants voted in the order Ballot A / Ballot B; the other half in the order Ballot B / Ballot A. For preference data, after voting both ballots, participants commented on 16 pairs of pages, giving us preference page by page, as well as an overall preference at the end.

Results

- Participants voted more accurately on the ballot with plain language instructions. (See full report, page 30, especially Table 3 on page 31.)

- Participants who voted the plain language ballot first (order B, then A) did significantly better on the traditional language ballot than participants who voted the traditional language ballot first (order A, then B). Working with Ballot A (traditional language) first did not help participants nearly as much in working secondly with Ballot B (plain language). (See full report, page 32, especially Figure 1 on page 33.)

- Education was significantly associated with errors: lower education – more errors. That association was more pronounced with the traditional language ballot than with the plain language ballot. (See full report, page 33, especially Table 4 on page 34.)

- Participants could tell the difference in the language of the two ballots. When doing a direct comparison of 16 specific pages from the two ballots, participants preferred the plain language ballot by a very wide margin on 12 of those 16 pages. The wide margins ranged from 64% to 98%. (See full report, page 34-36.)

- Asked for an overall preference, participants overwhelmingly chose the plain language ballot (82%). (See full report, page 36.)

What we learned

Participants voted more accurately on the ballot with plain language instructions (See full report, page 30, especially Table 3, page 31)

- Participants voted more accurately on Ballot B (plain language) than on Ballot A (traditional language).
- Ballot A mean accuracy 15.5; Ballot B mean accuracy 16.1; This difference was found to be marginally statistically significant using within-subjects analysis of variance (ANOVA) ($F_{1,43}$=3.413, $p < .071$).

Participants who voted B first did better on A than participants who voted A first (See full report, page 32, especially Figure 1, page 33)

- Working with the plain language ballot (B) first helped participants do better on the traditional language ballot (A). Working with Ballot A first did not help participants nearly as much in working secondly with Ballot B.
- Accuracy on Ballot A increased from 14.4 to 16.3 when it followed Ballot B; statistically significant using within-subjects ANOVA ($F_{1,43}$=23.057, $p < .001$).

Lower education was associated with more errors
(See full report, page 33, especially Table 4, page 34)

- Geographic location, gender, age, and voting experience were not statistically significant differentiators of accuracy.
- Education was statistically significant. Participants with less education made more errors; ($r = -.419$, $p < .004$, effect size $R^2 = 0.176$).

Education made a slightly greater difference for Ballot A than for Ballot B (See full report, page 34)

- The correlation of lower education with more errors was slightly stronger with Ballot A – traditional language ($r = -.393$, $p < .008$, $R^2 = 0.154$) than it was with Ballot B – plain language ($r = -.359$, $p < .015$, $R^2 = 0.129$).
- A within-subjects ANOVA, however, revealed that the difference between the impact of education on accuracy for Ballot A and the impact of education on accuracy for Ballot B, while a trend, was not statistically significant ($F4,40 = 1.114$, $p < .364$).

Participants could tell the difference in the language of the two ballots (See full report, pages 34-36)

- When doing a direct comparison of 16 specific pages from the two ballots, participants preferred the plain language ballot by a very wide margin on 12 of those 16 pages. The margins ranged from 64% to 98%.

- In the interview, the moderator only said, "Notice that the instructions on these pages are different. Please compare them and comment on them." The moderator did not mention "plain language" or explain anything beyond that one sentence. Participants' comments were almost always in terms of the features of the instructions that follow from plain language guidelines. Their comments indicated that they could tell the difference and that they preferred instructions developed according to plain language guidelines.

Participants overwhelmingly preferred the plain language instructions (See full report, page 36)

- 82% (37 of 45 participants) chose Ballot B for their overall preference.
- 9% (4 of 45) chose Ballot A.
- 9% (4 of 45) chose "no preference."
- The choice of the plain language instructions for ballots was statistically significant (p<.001).

Where (and why) participants had problems

To understand the problems participants had, we used our notes and videos to add data from observations to the information from the voting results files that were the basis of the accuracy results.

Eight pages caused participants serious problems

Six pages had error rates of 13.3% or higher (6 participants or more): Straight Party Vote/Straight Party Voting, US Senate, Registrar of Deeds, State Senator, County Commissioners, City Council. In addition, our observations tell us that two other pages were very problematic: the second Straight Party Vote/Straight Party Voting page on which participants chose whether to review or bypass/skip the partisan contests; the Ballot Summary (A)/Review Your Choices (B) page that showed how the participant had voted.

Straight-party voting is a difficult concept for many voters and changing a party-based choice is an even more difficult concept

Many of our participants were not familiar with the option to vote a straight-party ticket. Even more were confused by the idea of voting straight-party and then changing a party-based choice.

On a paper ballot with straight-party voting, there are only two options: vote straight-party without marking votes in individual party-based contests or vote only in individual contests. Electronic voting provides the option for voters to change their votes in individual contests after selecting to vote a straight-party ticket. But for many of our participants, there was a conflict in logic trying to understand the possibility of voting straight-party and then being able to change some of those votes.

Plain language helped somewhat with explaining the difficult concept of straight-party voting and changing a party-based choice. More participants made errors on A than on B related to this problem. The instructions on the plain language ballot were longer and more informative – and that helped participants.

Some participants were not clear or confident about the difference between US Senate and State Senator

The US Senate contest came earlier in the ballot than the State Senator contest. The direction we gave participants was to change their choice for State Senator, but 9 participants on A and 7 participants on B changed the US Senate contest instead of (or in addition to) changing the State Senator contest – and did not go back to fix that problem even if they later realized they had changed the wrong contest.

Plain language was only a factor in that it helped people understand how to change a vote. Language did not otherwise differentiate the ballots related to this problem.

The progressive disclosure of an electronic ballot hinders voters from answering their questions about which contest is which

On an electronic ballot, voters see only one contest at a time and have no information about the choices that come later on the ballot. When they came to the contest for US Senate, they did not know that there was another Senate contest further on in the ballot. That participants made fewer of these errors on the second ballot they voted attests to the progressive disclosure (not seeing what's yet to come) as part of the problem.

Deselecting is a difficult concept

When they got to the first place where they needed to change a vote (for many this was on the Registrar of Deeds page), many participants did not remember that they had to deselect the already-selected candidate. They tapped (touched) several times on the one they now wanted without first tapping on the one they did not want.

Participants told us that the instructions on how to vote and change a vote were much clearer on the plain language ballot. The traditional language ballot had instructions in one dense paragraph with no highlighting. The plain language ballot had instructions in three short paragraphs with space between and bold highlighting of key words.

However, on Registrar of Deeds, participants had to write in a candidate. When they could not get deselecting/selecting to work and went back to the instructions, they misdiagnosed their problem. They read the paragraph about writing in a candidate and did not go on to read the paragraph about changing a vote (their real problem). In the recommendation section, we recommend changing the order of those instructions.

Some participants confused the County Commissioner contest with the City Council Contest

Our directions did not tell participants to make any changes to their straight-party votes for County Commissioner. Participants were told to make changes in the City Council contest. The County Commissioner contest came earlier in the ballot than the City Council contest. In a problem similar to that of US Senate and State Senator, the high error rate on the County Commissioner page is largely due to participants not realizing or not paying attention to the difference between the County contest and the City contest.

Showing undervotes in red on the Summary/Review page caused some participants to insist on voting until the red disappeared

From our observations, 17 participants (37.8%) asked questions, expressed concerns, and in several cases actually changed votes because the red color bothered them so much. Some added votes for people they did not particularly want just to be sure their votes for the other candidates would count. A few wrote in candidates to fill the remaining votes. One even cast blank write-ins. All these actions were to get the red blocks to turn blue – signaling a complete vote.

Participants told us that the information was much clearer on the plain language Review Your Choices page than on the traditional language Ballot Summary page. They were much better at changing votes with the clear step-by-step instructions of B. However, some still saw the red blocks as sending the message "you must vote until the red is gone," even though that's not what the instructions on the page said.

For more details on problems voters had, see the full report, Part 3, pages 40 – 71.

For details of participants' preferences and their comments comparing the language of the two ballots, see the full report, Part 4, pages 72 – 102.

Recommendations

Use plain language in instructions to voters

The success of Ballot B in both performance and preference strongly supports a recommendation that all ballots should follow plain language guidelines. In the full report (on pages 103 – 105), we list the guidelines that we followed in creating the plain language ballot, including guidelines on what to say, where to say it, how to say it, and what to make it look like.

Use Ballot B language with specified changes

Most of the specific language in Ballot B worked very well and produced the strong performance and preference differences that we saw in the results. We would, however, change a few specific elements to go even further in plain language. We give those in the full report on pages 105 - 107. An important change we recommend is to give instructions for how to deselect an already-selected choice on every page where that action might be needed.

Put each contest and measure on its own page on a Direct Recording Electronic (DRE) voting system

Another critical guideline that we followed with both ballots was to present each contest and measure on its own page. Other research has shown that this is vital. Many people miss the second race when one DRE screen has more than one race.

Consider removing straight-party options from ballots

This study showed that much as plain language can help, it cannot solve all problems in voting. Most of the errors that our participants made were related to straight-party voting and wanting to change some party-based races after voting straight party.

We recommend that states think about this issue and review their policies with consideration of our findings.

Do more with voter education materials

In addition to their problems understanding straight party, many of our participants did not have a clear concept of the different levels of government. They mistook the U S Senate race for the State Senator race. They mistook the County Commissioner race for the City Council race.

The language on the ballot itself cannot compensate for this lack of understanding. Voter education before the election is needed. To be successful, voter education materials must also be in plain language, and the guidance in these recommendations for language and layout is relevant to all materials for voters.

Furthermore, sample ballots must look like the ballots that voters will use in the polling place.

Test ballots with voters before each election

Based on this study, we can strongly recommend the design and language of Ballot B for all ballots (with the changes listed in the recommendations in the full report). However, no specific ballot for any specific election in any specific jurisdiction is going to have exactly the contests and measures that we included in Ballot B. Local election officials constructing ballots are going to continue to make choices and decisions on every page of every ballot whether the ballot is delivered on paper or electronically.

The best way to guard against disaster in an election due to ballot design or language is to have a few voters try out the ballot before the design and language become final. The methodology for having voters try out a draft is usability testing. The participants in these usability tests must be voters from the community, not people who work in the Election Department. People who work in the Election Department, even if they do not work directly in designing or defining the ballot, are likely to know more about voting and more voting vocabulary than the typical voter.

For more details on recommendations, see the full report, Part 5, pages103 – 109.

Suggestions for future research

Test with low literacy voters

In this study, participants with lower education levels made significantly more errors (particularly on the traditional language ballot). But we did not specifically test for low literacy; and our participants, even those with lower education levels, read competently. If the low-education-level but fairly competent readers in our study had problems, readers with lower reading skills might have even more problems. Further research with low literacy voters would be worthwhile.

Investigate the prevalence of people who vote empty ballots

We actually conducted 46 sessions. We analyzed data from only 45 of those sessions because one participant voted an entirely empty ballot for both A and B. She met our screening criteria and, indeed, reported that she is a high school graduate. She read our directions and reported that she had voted successfully on both ballots. But she moved through them only clicking on Next and never actually selecting a party, candidate, or option on a measure. Although we had only one participant who did this, we wonder how often empty ballots are cast in real elections. How large a population of eligible voters does this person represent? What can be done to help this person and others like her cast votes that match her real intent?

Test with older adults

In our study with participants ranging in age from 18 to 61, we did not find differences in voting behavior among the different age groups. We did not have anyone older than 61 among our participants. We would be very comfortable making the hypothesis that plain language would matter as much if not more to older adults, but further research with that population would be needed to test that hypothesis.

Test with other modalities (for example, audio) and with special populations

Our study did not focus on modalities other than text nor on people with specific needs. As we just said about older adults, we would be very comfortable making the hypothesis that plain language would work better than traditional language for voters who listen to instructions (audio) and for voters with cognitive issues. However, further research would be needed with those populations to test our hypothesis.

Test with other languages

Plain language is not just an English-language issue. Jurisdictions that prepare ballots in languages other than English must consider the value of applying similar guidelines for those languages. We believe that clear, simple, direct, specific wording and presentation of instructions helps all voters in all languages.

It is a known fact in other domains (translating manuals, for example) that plain language facilitates translation. For jurisdictions that start in English and translate into other languages, a plain language ballot in English should make translation faster, easier, and more accurate. We suggest replicating this study with other languages. We suggest that jurisdictions that translate from English keep track of costs of translation for traditional, non-plain-language ballots compared to translation for plain-language ballots.

Apply what we learned to paper ballots

We studied the language for the pages of a touch screen ballot. Much of what we learned applies to paper ballots as well. Paper ballots, however, operate differently both in terms of what a voter can do and in constraints on where and how instructions can appear. A study comparable to the one we completed should be done with paper ballots.

Do a similar study on a ballot without a straight-party option

Many of the problems we saw came from our participants not understanding the concept of straight-party and especially of being able to both vote straight-party and change a party-based contest. How many of the errors that we saw even with our plain language Ballot B would go away if straight-party were not an option? Our hypothesis would be that voters would make fewer errors even though they would have more contests to go through; but that is an empirical question.

Find the best way to create and deliver voter education materials

Many people, besides our participant who voted two empty ballots, showed that they had little concept of voting and the many types of contests in elections. What type of voter education is most effective in helping these people understand the process of voting, where races come on ballots, what the different levels of government are, etc.?

Look into changes for specific issues that came up in this study

Although the plain language Ballot B was much more effective in helping voters than the traditional language Ballot A, people still had problems with some aspects both of the ballot and of voting. Three examples of specific studies that would be worthwhile:

- **Deselecting** was not a natural behavior for many of our participants. They did not remember that to change a vote in a contest where the maximum number of people or options was already selected they had to first click on a candidate or option that they no longer wanted. Their natural instinct was to simply chose the new candidate or new option. We recommend further study of this issue.

- **Red as the color to indicate undervoting** made several of our participants so uneasy that they insisted on voting for the maximum number to remove them. We have suggested a different scheme for the Summary/Review page. Would our suggested scheme of red/orange/green solve the problem? Would some other way of indicating that intentional undervoting is okay work better than our suggestion? For example, would it help voters if we put a comment on every ballot page saying that "You may vote for fewer than x candidates"?

- This study resolved only some of the issues **concerning the best way to tell people how many candidates they could vote for** in a contest.

  This study may have resolved the question: Does the instruction have to mention a possibility of voting for none? Participants did not have a problem voting for none based on our directions. When we specifically told people to leave a contest unvoted, they did that. (We later had them go back and vote the unvoted contest.)

  This study found a successful way to give an instruction in a single-candidate contest. We did not vary this instruction between the ballots; both had Vote for one. None of our participants had a problem with the instruction. However, we do not know if giving that instruction with the numeral would have been better: Vote for 1. A study to test that difference would be useful.

  This study did not resolve the issue of how to best express the instruction in a multi-candidate contest. Slightly more participants preferred the formula, Vote for no more than x to the formula, Vote for [each number listed out] in contests with four and five as the maximum. However, slightly more participants preferred the numbers spelled out when the maximum was two.

Add other aspects of the voting process that we did not include

Our ballots did not give participants a paper trail. Just having paper is not in itself enough to ensure that people will notice the paper, be able to read it and review it before casting their vote, understand what to do if they do not agree with the paper, and so on.

The design, language, and usability of audit screens and audit paper are also important issues and should be researched.

What we did

Research questions

- Do voters vote more accurately on a ballot with plain language instructions than on a ballot with traditional instructions?
- Do voters recognize the difference in language between the two ballots?
- Do voters prefer one ballot over the other?

Participants

- 45 people in 3 locations (Georgia, Maryland, Michigan)
- Ages: 18 to 61; average age 36
- Education: Mostly less than high school, high school, some college
- All US citizens eligible to vote but may not have ever voted
- English speaking, but not necessarily native speakers

Sessions

- May and June 2008
- Individual sessions of about one hour
- Performance: Voting on two ballots (either in order A, B or in order B, A)
- Preference: Looking at 16 sets of comparable pages from the two ballots, participant discussed whatever differences he or she saw, stated preference for which was "best for a ballot," and gave written overall preference for A, B, or no preference
- Demographic/voting/technology questionnaire: Participant answered a short questionnaire about self, voting experience, and experience with other technology

Materials

- Two ballots that differed only in the wording and placement of instructions and the names of candidates and wording of measures.
- Screen design of both ballots was the same.
- Ballot A used traditional language taken from contemporary ballots.
- Ballot B used plain language based on best practices in giving instructions.
- Other materials included: screener for recruiting, script for moderator, informed consent form, directions to participants on the candidates and measures to vote for, questionnaires (preference; demographic/voting/technology), note-taking forms.

Presentation of ballots

- Set up to be like current Direct Recording Electronic (DRE) voting systems.

- Ballots programmed into and presented on Sahara Slate Tablet PCs.

Ballots and tasks participants did on the ballots

- Both ballots included a straight-party option, 10 party-based contests, 4 non-party-based contests, and 3 amendments/measures.

- Both ballots also included instructions on how to vote, a screen for writing in candidates, a Ballot Summary (A)/Review Your Choices (B) screen at the end, a Confirm screen, and a Thank You screen.

- Participants received and read a list of directions before voting each ballot. The lists for Ballot A and Ballot B were identical except for the names of people for whom participants were to vote. Participants kept the directions with them while voting and could refer to the directions as they voted.

- If participants followed all the directions, they did these tasks:
  - Voted for all the candidates from one party at the same time (straight-party).
  - Reviewed the straight-party candidates to accomplish some of the other directions.
  - Did not change many of the straight-party candidates as they reviewed the straight-party votes.
  - Wrote in a candidate instead of their party's candidate.
  - Changed a vote from the candidate of their party to the candidate of another party in a "vote for one" contest.
  - Changed votes in a "vote for no more than four" contest. (This and the previous two tasks required "deselecting" one or more of their party's candidates if they had successfully voted straight-party.)
  - Skipped a contest.
  - Voted per the directions in several non-party-based contests and for three amendments/measures.
  - Went back and voted the skipped contest from the Summary/Review page.
  - Changed a vote from the Summary/Review page. (This and the previous task were directions given on paper to the participant at the appropriate time – when the participant was on the Summary/Review page.)
  - Cast the ballot and confirmed the casting.

- Note that the directions participants saw never used the words "straight-party," "partisan," "non-partisan," or "write in." We couched each direction in a sentence that put participants into the voting role. For example, the direction for the task of writing in a candidate for Ballot A was:

  > Even though you voted for everyone in the Tan party, for Registrar of Deeds, you want Herbert Liddicoat. Vote for him.

  When they got to the Registrar for Deeds contest, participants saw that Herbert Liddicoat was not on the ballot. They then had to 1) realize that they needed to write him in and 2) succeed at writing his name in the right way.

For more details on what we did, see the full report, Part 1, pages 13 – 29.

Report of Findings:
Language in Ballot Instructions

Part 1: Description of the Study

What did we study?

In this study, we compared two ballots that differed only in the wording and presentation of the language on the ballots presented to the voter. For Ballot A, we used conventional wording and presentation, taken from typical ballots. For Ballot B, we used plain language wording and presentation.

What is plain language?

A document is in plain language when the users of that document can quickly and easily find what they need, understand what they find, and act appropriately on that understanding. For more details, examples, and resources about plain language, see http://www.plainlanguage.gov/.

For ballots, eight of the most critical plain language guidelines are the following:

- Be specific. Give the information people need.

- Break information into short sections that each cover only one point.

- Write short sentences.

- Use short, simple, everyday words.

- Address the reader directly with "you" or the imperative ("Do x.")

- Write in the active voice, where the person doing the action comes before the verb.

- Write in the positive. Tell people what to do rather than what not to do.

- Put context before action, "if" before "then."

For a more detailed list of the plain language guidelines that we used in this study, see the section, "Recommendation 1. Use plain language in instructions to voters" in Part 5,

Recommendations, later in this report. We have also pulled out the complete list of plain language guidelines and made it one of the appendices.

You can see how we applied the guidelines to the ballots by looking at Ballot A (traditional language) and Ballot B (plain language) in the appendices.

Although this study was limited to comparing two ballots in English, the issue of the wording and presentation of language on ballots is relevant to other languages as well. That is why we refer to "plain language" and not to "plain English." In Part 6, Conclusions and suggestions for future research, we recommend conducting similar studies on ballots in other languages.

What were the goals of the study?

We set up this study to answer three questions:

- Do voters vote more accurately on a ballot with plain language instructions than on a ballot with traditional instructions?

- Do voters recognize the difference in language between the two ballots?

- Do voters prefer one ballot over the other?

Where did the traditional and plain language instructions come from?

We based the traditional version of the ballot instructions on typical features of contemporary ballots. We based the plain language version of the ballot instructions on best practices in giving instructions.

In previous work for NIST, Ginny Redish, a linguist and plain language expert, reviewed more than 100 ballots from all 50 states and the District of Columbia. This gave us the traditional language for Ballot A.

In that project, Dr. Redish also analyzed the gap between the instructions on those ballots and best practices in giving instructions (report available at http://www.vote.nist.gov/instructiongap.pdf). Dr. Redish then developed a set of guidelines for writing clear instructions for voters, focusing on the issues that arose in her earlier analysis (guidelines available at http://www.vote.nist.gov/032906PlainLanguageRpt.pdf). This work gave us the plain language guidelines for Ballot B.[1]

How did we conduct the study?

In the following sections we describe the methodology in detail by answering these questions:

How many people participated in the study?

When and where did we conduct the study?

Who participated?

How did we find the participants?

What were the ballots like for this study?

What did we do in each session?

How were participants compensated?

What was the setup like in each location?

How were the ballots presented?

What materials did we use?

How did we control for bias in the study?

What tasks and directions did we give the participants as voters?

What data did we collect while participants voted?

What data did we collect in the forced choice comparison interview?

---

1 For other sources of plain language guidelines, see:

GMAP (Government Management Accountability & Performance), *General guidelines*, Plain Talk, http://www.accountability.wa.gov/plaintalk/ptguidelines/default.asp.

Office of the Federal Register, *Drafting Legal Documents*, http://www.archives.gov/federal-register/write/legal-docs/clear-writing.html.

www.plainlanguage.gov; especially http://www.plainlanguage.gov/howto/guidelines/bigdoc/TOC.cfm.

Securities and Exchange Commission, *A plain English handbook: How to create clear SEC disclosure documents*, http://www.sec.gov/pdf/handbook.pdf

How many people participated in the study?

After piloting the study with 6 participants in one location (Baltimore, Maryland) and finalizing the materials and study plan based on the pilot, we collected data with 46 participants in three locations.

One participant was such an "outlier" that we did not include data from that session in our analysis. This participant voted an entirely empty ballot for both ballots. She fit our screening criteria; in fact, she was American-born, a native English speaker in her 50s, and reported that she was a high school graduate. She went through each ballot, clicking Next to move from screen to screen, but not selecting any candidate or option. When asked if she was voting according to our directions (which she had read and which she had with her), she said yes, she was. At the end of each ballot, despite the red boxes on the Summary/Review screen, she cast the ballot. When asked how she thought she had done, she told us she believed that she had voted successfully.

Because her session was the only one in which this happened, we consider it an "outlier" situation for this study. However, this participant may represent a portion of the voting population that deserves further study. We raise that issue in Part 6: Conclusions and suggestions for future research.

This report, therefore, is based on 45 participants who worked with the final ballots.

When and where did we conduct the test sessions?

We conducted the test sessions in May and June, 2008. Our three locations (in alphabetical order) were

- Baltimore, Maryland

- East Lansing, Michigan

- Marietta, Georgia

We chose those locations for both geographic spread (Middle Atlantic, South, Midwest) and diversity in the type of community (urban, small town, suburban community with a large minority population). Dividing our 45 participants among three sites gave us enough participants in each location to ask whether location affected the results. As you will see in Part 2, Results, location did not affect the results. In each location, we held the sessions in the usability center of a university:

- University of Baltimore in Baltimore, Maryland

- Michigan State University in East Lansing, Michigan

- Southern Polytechnic State University in Marietta, Georgia

However, our participants were not students at those universities. They were people who live or work in the local communities. (Some of our participants were taking college classes; but, in fact, they were not studying at the institution where they came to participate in the study.)

Going to those locations worked well for the study. The usability centers provided us with the support we needed: someone to greet participants, a comfortable place for participants to wait if they were early, a location that people could find easily and were comfortable coming to, pleasant facilities to work in, and technical support when we needed it.

How did we recruit our participants?

Based on best practices in usability studies (Dumas and Redish, 1999, chapter 10; Rubin and Chisnell, 2008, chapter 7),[2] we recruited a diverse set of participants, drawn from three geographic areas and focusing on lower education levels, to get sufficient data for analysis.

We recruited with only two screening criteria

We recruited based on these two criteria:

- American citizens 18 and older (that is, people who are eligible to vote, whether or not they have ever voted, whether or not they have ever registered to vote)

- fluent English speaking (as found in a telephone screening interview, so not necessarily native speakers)

All of our participants met these criteria.

---

[2] Dumas, J. S. and Redish, J. C., *A Practical Guide to Usability Testing*, Revised Edition, Intellect, 1999.

Rubin, J. and Chisnell, D., *Handbook of Usability Testing*, Second Edition, Wiley, 2008.

We strove for diversity

Although these were not screening criteria, we wanted at least some diversity in gender, ethnicity, and age.

- We tried to balance for gender and had 23 women and 22 men.

- We did not select for ethnicity or race but we did end up with a diverse set of participants. By our observation, we had 21 Caucasians and 24 people of other ancestry.

- We wanted people over a wide range of age. Our youngest participants were 18 years old; the oldest was 61. The average age was 36.

Given the size of our set of participants, we believe this is a good diversity of age. In Part 6, recommendations for future research, we recommend a study concentrating on senior citizens.

We focused on people with a high school education

Because we are concerned that ballots be understandable and usable to people regardless of their education, our study plan was to focus on people with high school or less or with some college but not advanced degrees. By including people with lower levels of education, we hoped to gain some understanding of issues that other researchers had raised about higher residual voting rates for voters with lower education levels.[3] In addition, we knew from our pilot study and research done by others that people with higher education levels typically have little trouble using ballots. ("Residual

3 Herrnson, Paul S., Richard G. Niemi, Michael J. Hanmer, Peter L. Francia, Benjamin B. Bederson, Frederick G. Conrad, and Michael Traugott, Voters' Evaluations of Electronic Voting Systems: Results from a Usability Field Study, *American Politics Research*, 36 (4), 580-611, 2008.

Norden, Lawrence. Jeremy Creelan, David Kimball, and Whitney Quesenbery, *The Machinery of Democracy: Usability of Voting Systems*, Brennan Center for Justice at NYU School of Law, 2006. Available at http://www.brennancenter.org/programs/ downloads/Usability8-28.pdf

Norden, Lawrence, David Kimball, Whitney Quesenbery, and Margaret Chen, *Better Ballots*, Brennan Center for Justice at NYU School of Law, 2008. Available at http://www.brennancenter.org/content/resource/better_ballots/

voting rates" means how often, in a particular contest, people do not vote or vote for fewer candidates than they could vote for – whether they intended to do that or not.)

Table 1 shows our participants by education level.

Table 1. Number of participants at each education level (N=45)

Highest education level achieved	Number of participants
Less than high school	9
High school graduate or GED	15
Some college or associates degree	12
Bachelor degree	8
Some courses beyond college	1

GED = General Education Development, a series of tests that people can take to show they have the equivalent of a high school education. Many people who drop out of high school take the GED later in life.

We hoped for a range of voting experience

In recruiting, we asked no questions about people's voting experience. Therefore, we did not know beforehand what diversity we would have among our participants in whether they had ever voted, how often they had voted, or the types of ballots they had used. We did ask about those experiences at the end of each session. As it turns out, just half of our participants had voted with a touch screen interface (23 of 45). Most (38 of 45) had used an automated teller machine (ATM) or bank machine, although not all ATMs are touch screen interfaces.

We looked at the correlation of experience voting with a touch-screen interface and performance on the ballots. Accuracy scores for people with and without previous touch-screen voting experience on Ballot A were almost identical. On Ballot B, the 22 people *without* previous touch-screen voting experience actually did a little better than those with touch-screen voting experience. However, that difference was not statistically significant. Previous voting experience with touch-screen interfaces did not correlate significantly with how participants performed in this study.

A professional recruiter helped us

A professional recruiter who recruits for studies like this in many locations across the country helped us find appropriate participants. They came to us through these channels:

- Community groups in the locations where we were testing

- Professional and personal networks

- Online classifieds

- Asking people who came through any of the first three channels to refer others who met the screening criteria.

Our recruiter conducted a search for people who met our requirements in each location (rather than working from a database). Using networking to find participants helped pre-qualify respondents, as people referred one another. As she gathered names, our recruiter contacted respondents by email first to determine whether they did indeed qualify for the study. Then she talked to potential participants by phone, finding out their availability, getting any further contact information necessary, and gathering appropriate referrals to other potential participants.

Some of our participants, therefore, came to us because they responded to a request online. However, not all did. Some came through referrals. For example, one older gentleman had no email address. His niece read about the study and served as our initial contact to him. Furthermore, even though most of our participants used email, had a cell phone, and were savvy about other technology, their sophistication with technology did not necessarily mean that they understood what a ballot is like, were used to ballots, or could vote accurately.

A few participants (including the participant whose session we did not include in the analysis) were recruited on the spot to fill empty slots or to substitute for a "no-show." In every case, they met the screening criteria.

For details on our participants' demographic characteristics, voting experience, and experience with technology, see the appendices.

If the respondents did not qualify for the study, the recruiter destroyed any personal information she had for them, such as phone numbers and email addresses.

What were the ballots like for this study?

In this study, we compared two ballots that differed only in

- the wording and placement of instructions

- the names of candidates and the wording of measures

We adapted our ballots from the NIST "medium" ballot developed by the company, User-Centered Design.[4] The medium ballot from User-Centered Design was also used by Design for Democracy/AIGA in its project for the Election Assistance Commission.[5]

The NIST medium ballot, as developed by User-Centered Design, includes straight-party voting, and has 12 party-based contests, two retention questions, and six referenda. In some contests, it has more than one screen of candidates.

We adapted this ballot by reducing slightly the number of party-based contests and the number of referenda, including a few non-party based contests, and never having more than one screen of candidates in a contest. The ballots in our study included straight-party voting, nine party-based contests, three non-party-based contests, two retention questions, and four referenda.

The screen design and navigation were identical for both of our ballots.

We kept the same type font and type size in both ballots. We also followed best practices in information design. For example, although many ballots today still use all capital letters for instructions, we know from research on design that all capitals are more difficult to read than appropriately-used upper- and lower-case. Both of our ballots were entirely in appropriate upper- and lower-case.

The political parties were indicated by color names to avoid any bias for or against actual parties. We did not name any party either Red or Blue. Candidates' names were made up but resembled a range of typical names. Research by the ACCURATE group at Rice University has shown that study participants are just as accurate and not put off by

---

4 See the report from User-Centered Design to NIST, "Preliminary Report on the Development of a User-Based Conformance Test for the Usability of Voting Equipment," dated March 10, 2006, The report, which includes logical specifications for the ballot, is available at http://vote.nist.gov/032906User-BasedConfTesting3-10-06.doc

5 The Design for Democracy/AIGA report is available at http://www.aiga.org/content.cfm/design-for-democracy-eac-reports

voting ballots with made-up names as by voting ballots with names of people they recognize.[6]

The two ballots are in the appendices. The ballot with more traditional instructions is Ballot A. The ballot with more plain language instructions is Ballot B.

What did we do in each session?

Each participant came for an individual one-hour session.

The sessions had these major parts:

- Introduction and signing Informed Consent Form

- Voting on two ballots in sequence (A, B or B, A) conducted in typical usability testing fashion. (Typical usability testing fashion means, one participant at a time actually working with each ballot while voicing all thoughts [think aloud]. We recorded what was happening on the screen and what the participant was saying with video / audio recording software, while the data collector observed and took notes.

- Forced-choice page-by-page comparison of 16 pages of the two ballots with a written request for a final overall preference

- Questionnaire about demographics, voting experience, and experience with technology, followed by our thanking the participant and giving the incentive payment

Table 2 shows what we did in each session in more detail. The timing of actual sessions ranged from about 45 minutes to about 70 minutes.

Table 2. Details of the activities in each session

Item	Description of activities in that part of the session
Overview of session	Using a script, the session moderator described the objectives of the session, her role, the role of the note-taker, and the participant's role, and explained the Informed Consent Form.

6 Everett, Sarah, Michael Byrne, and Kristen Greene, Measuring the Usability of Paper Ballots: Efficiency, Effectiveness, and Satisfaction, in Proceedings of the Human Factors and Ergonomics Society 50th Annual Meeting. Santa Monica, CA: Human Factors and Ergonomics Society, 2006. Available at chil.rice.edu/research/pdf/EverettByrneG_06.pdf

Item	Description of activities in that part of the session
Consent Form	The participant read and signed the Informed Consent Form.
Vote ballot 1 (A or B depending on participant's order)	Working from directions for that specific ballot provided by the moderator, the participant voted on the first ballot. Participants were asked to read through the directions for that specific ballot before beginning to vote. They kept the directions with them to refer to as they went through the ballot. Participants with odd numbers (A1, A3, and so on) voted Ballot A first. Participants with even numbers (A2, A4, and so on) voted Ballot B first.
Subjective ratings for first ballot	When the participant finished voting on the first ballot, the moderator asked the participant to rate 5 statements on a scale ranging from 1, "Strongly disagree" to 5, "Strongly agree."
Vote ballot 2 (A or B depending on participant's order)	Working from directions for that specific ballot provided by the moderator, the participant voted on the second ballot. The lists of directions were identical for the two ballots, differing only in the colors of the parties and the names of the candidates the participant was told to vote for. As with the first ballot, participants read through the directions before beginning to vote and kept the directions with them to refer to as they went through the ballot.
Subjective ratings for second ballot	When the participant finished voting on the second ballot, the moderator asked the participant to rate 5 statements using a scale from 1, "Strongly disagree" to 5, "Strongly agree" on the same sheet used for the first ballot. Participants could change their ratings for the first ballot.
Comparative interview	Showing paper printouts of each page type for each ballot side by side, the moderator asked participants to comment on the differences between the two versions of each page and then select one version as the "best for a ballot." We audio-taped this part of the session.
Preference questionnaire	Participants selected the ballot version they preferred overall and wrote out their reasons for their preference. They had three choices: A, B, no preference.
Final questionnaire and ending	Participants answered 8 multiple-choice questions about themselves, their voting experience, and technology they use regularly. We thanked the participants, gave them the incentive, and walked them out of the study area.

How were participants compensated?

Each participant was paid $75 in cash at the end of the session.

What was the setup like in each location?

Although the layout of the room we worked in was different in each location, we were able to set up a similar situation in each location. We had three "stations" in one room. The photos below show the "stations" for our sessions in Baltimore.

Figure 1. The session moderator at Station 1. The participant sat in the other chair (pulled up to the table) for the introduction and to sign the Informed Consent Form. The moderator and the participant returned to this table for the interview in which the participant compared ballot pages.

Figure 2. Stations 2 and 3, showing the touch-screen tablet PCs on which the ballots were presented. The participant sat in front of the tablet PC with the moderator next to the participant and the note-taker behind. One station was for Ballot A; the other for Ballot B. The participant and the data collection team moved from one station to the other so the participant could vote on both ballots. (Which station we started at depended on whether the participant was voting in the order A, B or the order B, A.)

How were the ballots presented?

The ballots were programmed into and presented on identical touch-screen tablet PCs. The PCs were Sahara Slate Tablekiosk L2500s with a 12.1" XGA screen.

We did not use any of the currently existing Direct Recording Electronic (DRE) voting systems for several reasons. We did not want to bias the study with the experience or lack of experience any participant might have had with one or another of the currently existing DREs. Because we were testing instructions and not navigation or casting modes, we did not want to test the specific modes or buttons of just one current DRE at the expense of not testing the modes or buttons of other DREs.

We used the two-column ballot design recommended by Design for Democracy. The ballot had instructions for a particular contest in the left column of the screen and candidates' names or other choices for that contest in the right column. It is not easy to program this type of ballot into the currently-existing DREs.

What materials did we use?

We developed several instruments to help us gather and organize data for this study:

- Screener to recruit participants

- Two ballots

- Script for moderator to talk to participants

- Informed Consent Form for participants to sign

- Directions to the participant on which party, candidates, and measures to vote for on each ballot

- Short satisfaction questionnaire for use after voting each ballot

- One-item questionnaire to indicate overall preference

- End-of-session questionnaire on demographics, voting experience, and experience with technology

You can see all of these materials in the appendices.

In addition to those materials, we had note-taking forms for the data collector:

- Note-taking form for usability testing

- Note-taking form for comparative interview

And we had printed versions of 16 pages from each ballot for the comparative interview.

How did we control for bias in the study?

For this study with 45 participants, each participant voted both ballots – a "within-subject" study.

(In a between-subjects study, each participant would work with only one of the two ballots. However, we would not then know the effect of differences in demographic factors or personality traits on differences in performance or understanding of the ballots.

With a between-subjects study, you must have large enough numbers of participants to outweigh the effects of differences between individual participants.)

The caveat for a within-subjects study is that you must counterbalance the presentation of the materials to eliminate the "practice effect." That is, people inevitably learn from their first experience with one version of the material and that affects their performance with the second version of the material.

To neutralize practice effects, half of the participants used the two ballots in the order A, B; and the other half used them in the order B, A.

In sections below where we give comments from specific participants, you may want to know which participants voted in which order. Participants with an odd number voted Ballot A first, Ballot B second. Participants with an even number voted Ballot B first, Ballot A second.

We also looked at order effects in our data analysis. See Part 2, Results.

What tasks and directions did we give the participants as voters?

Our ballots included a straight-party option, 10 partisan (party-based) contests, 4 non-partisan (not party-based) contests, and 3 amendments/measures.

On a ballot with a straight-party option, the voter may vote once next to the party's name and have that automatically register a vote for all candidates of that party for all party-based (partisan) contests.

Just before they voted each ballot, we gave participants a sheet of specific directions to work with. This sheet told participants what party to vote for, what party-based contests to change, which contest to write in a candidate, and how to vote in all the non-party-based contests and for all the amendments/measures. Participants kept these directions with them to refer to as they went through the ballot. We also had two additional directions that we gave to participants when they were at the Summary/Review page.

We couched each direction in a sentence that put participants into the voting role. For example, the direction for the task of writing in a candidate for Ballot A was:

> Even though you voted for everyone in the Tan party, for Registrar of Deeds, you want Herbert Liddicoat. Vote for him.

When they got to the Registrar for Deeds contest, participants saw that Herbert Liddicoat was not on the ballot. They then had to 1) realize that they needed to write him in and 2) succeed at writing his name in the right way.

(See the bulleted list below for the tasks participants did, and see the appendices for the specific directions we gave to participants.)

Note that we did not use the terms "straight-party," "write in," "partisan," or "non-partisan" in our directions to voters. We did not use these terms because we did not want to lead or prime participants to look for key words that typical voters might not have in mind when they come to vote.

The following list recaps the tasks – the different voting behaviors – that we included in the study:

- Vote for all the candidates from one party at the same time (straight-party).

- Review the straight-party candidates to accomplish some of the other directions.

- Leave many of the straight-party candidates as they reviewed the straight-party votes.

- Write in a candidate instead of their party's candidate.

- Change a vote from the candidate of their party to the candidate of another party in a "vote for one" contest.

- Change votes in a "vote for no more than four" contest. (This and the previous two tasks required "deselecting" one or more of their party's candidates if they had successfully voted straight-party.)

- Skip a contest.

- Vote per the directions in several non-party-based contests and for three amendments/measures.

- Go back and vote the skipped contest from the Summary/Review page.

- Change a vote from the Summary/Review page. (This and the previous task were directions given on paper to the participant at the appropriate time – when the participant was on the Summary/Review page.)

- Cast the vote and confirm the casting.

What data did we collect while participants voted?

When the participant cast his or her vote on each ballot, the tablet PC automatically recorded that person's vote for each contest and amendment/measure. That gave us the accuracy of the vote. In this study, an error was casting a vote contrary to the directions we gave participants on who to vote for and how to vote on the amendments/measures.

As participants voted the two ballots, we collected observational data on where participants had problems using the ballot and errors they made in completing tasks we gave them. We also collected participants' comments that indicated specific understanding or misunderstanding of instructions.

In analyzing the data, we identified different types of errors that participants made on the ballots. We discuss those in Part 3, Discussion: Where did participants have problems?

What data did we collect in the forced choice comparison interview?

In the interview after participants voted both ballots, the moderator and the participant sat side by side. The moderator had two stacks of 16 pages – the same pages from both ballots. After explaining what we wanted, she turned over the first page of both ballots at the same time, always putting the ballot the participant voted first on the left side. Each page clearly indicated which ballot it was from, and the moderator reinforced that by pointing to the letter (A, B) on the first few pages as she turned them up.

The participant looked over the two comparable pages and commented on the instructions in any way that he or she wanted. When the participant finished all the comments he or she wanted to make, if the participant had not expressed a preference, the moderator pressed for a preference.

We then repeated that procedure with each of the other sets of comparable pages.

Using a prepared note-taking form, we collected data about which instructions the participant preferred for each of the 16 specific pages of the ballots where the instructions differ between Ballot A and Ballot B. The note-taker also wrote down the participants' comments on the instructions and reasons for the preference. We also audio-taped this part of the session as back-up to the data collector's notes.

After the moderator and participant had gone through the 16 pages, participants filled out a short questionnaire asking them which ballot they preferred overall and why. On this questionnaire, they could chose Ballot A, Ballot B, or No preference.

Part 2: Results

In this part, we report results for the three questions that we listed at the beginning of the report:

- Do voters vote more accurately on a ballot with plain language instructions than on a ballot with traditional instructions?

- Do voters recognize the difference in language between the two ballots?

- Do voters prefer one ballot over the other?

The answer to all three questions is "yes." Plain language makes a positive difference in both performance and preference.

Participants voted more accurately on the ballot with plain language instructions

Plain language instructions help voters vote more accurately.

What did we count as accuracy?

In this study, the success criterion – a correct vote – was a vote that matched the directions we gave participants on who to vote for and how to vote on the amendments/measures.

There were 18 pages in the ballot where participants voted (plus 8 other non-voting pages). We gave participants 12 explicit directions for voting on 11 of those 18 pages. There were 10 directions in the initial set plus two that sent participants back from the Summary/Review page to ballot pages they had previously been to (Water Commissioner and State Assembly). For 7 pages, we gave no directions, but another direction (to vote for a certain party at the straight-party option) automatically produced a correct vote on those pages. Therefore, there was an implicit direction to not change votes on those 7 pages.

Of the 12 directions, 10 were specific – the response was either correct or it was not. In two contests we gave participants leeway in completing their votes, but there were still specific correct votes. Those two contests were for City Council and Water Commissioner. Both were multi-candidate races.

On City Council, participants could vote for up to four candidates. The direction for that contest said that the voter must vote for the women, but that the voter could decide what to do with the rest of the votes for that contest. As long as participants voted for the two women candidates with or without other candidates, including writing in candidates, their responses were counted as correct. If participants did not vote for the two women candidates, their responses were incorrect.

For Water Commissioner, participants could vote for up to two candidates. There were only two available, along with options for writing in candidates. The direction for that contest the first time through the ballot was to skip the contest. After participants had reached the Summary/Review page, the moderator gave the additional direction, "You decide that you should vote for the Water Commissioners, so do that now." We did not tell participants how many people to vote for or that they were or were not allowed write-in candidates. As long as participants voted for at least one candidate – either on the ballot or as a write-in – their response was counted as correct. If they did not vote in the contest, their response was incorrect.

How did accuracy compare between the two ballots?

Table 3 shows the correct and incorrect votes on the two ballots: Ballot A with traditional language instructions; Ballot B with plain language instructions.

Table 3. Accuracy of votes on the two ballots for all participants
(45 participants, 18 possible correct votes on each of two ballots)

	Ballot A	Ballot B	Total
Correct	698	726	1424
Incorrect	112	84	196
Total	810	810	1620

We conducted a within-subjects (or repeated measures) analysis of variance (ANOVA) between the number of correct votes for the ballots.[7] (Ballot A mean of 15.5; Ballot B

---

7 The mean (or average) number of correct votes is an estimate of the typical response, which you get by totaling the scores for all participants in a group and then dividing by the number of participants in that group. An analysis of variance (ANOVA) is a test that statisticians use to see if the difference in responses for a particular factor being studied is

mean of 16.1). The difference in accuracy between the two ballots is marginally statistically significant ($F_{1,43}$=3.413, p < .071).[8] Participants voted more accurately on the ballot with plain language instructions.

Participants who voted B first did better on A than participants who voted A first

Working with Ballot B first helped participants do better on Ballot A. Working with Ballot A first did not help participants nearly as much in working with Ballot B.

As Figure 1 shows, there was little difference for the number of correct votes on Ballot B whether participants worked with it first or second. However, the number of correct votes on Ballot A increased from 14.4 to 16.3 when it followed Ballot B. This interaction between which ballot was seen first and the total number of correct items on a given ballot is statistically significant ($F_{1,43}$=23.057, p < .001).

This result suggests that what participants learned from Ballot B transferred to Ballot A. Using the plain language instructions first helped participants when they got to the ballot with traditional instructions. The reverse order effect (traditional instructions helping on the plain language ballot) was not nearly as strong.

---

statistically meaningful. "p" stands for "probability" and refers to how likely it is that you would have gotten the observed result by chance. A "p" value of 0.001 means that there is only one chance in a thousand that the observed result would have happened when the default assumptions about the results are true.

8 For our statistical colleagues: Mauchly's test for sphericity was conducted for each within-subjects ANOVA, as appropriate, and found to be non-significant. This indicates that the assumption of sphericity, which could impact our statistical results, appears to be reasonable for this data.

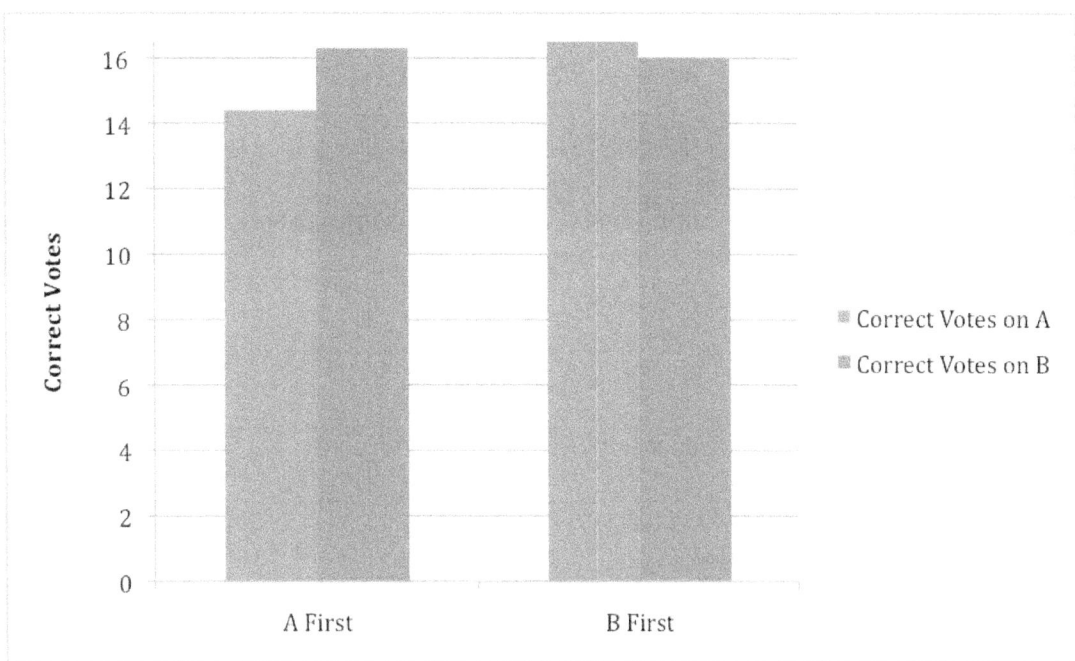

Figure 1. Participants who worked with B first (plain language ballot) did better on A (traditional language ballot) than participants who worked with A first.

Some participants spontaneously commented, either while they were working or in the interview later, on how B had helped them with A.

> B17 (getting to Ballot B after voting on Ballot A): "[B is] a little easier to read and understand. I like that."

> C32 (looking at the two instruction pages in the interview, referring to A):
> I started reading and it just became clear, maybe because I had already noted it on B.

> C42: (looking at the two instruction pages in the interview): B is better because there's a lot to remember. I learned from B.

Only education level made a difference in how accurately different groups of participants voted

We looked at correlations of accuracy with location (our three geographic sites) and with participants' characteristics (gender, age, voting experience, and education level). Location, gender, age, and voting experience were not statistically significant differentiators of accuracy. Note that our 45 participants ranged from 18 years old to 61 years old, with an average age of 36 years. As you can see in Volume 2, Appendix A, we had 7 participants 18 – 21 years old. Almost half of our participants (20) were 30 or younger – and age did not make a difference. However, we did not have participants in the 65+ age group. In Part 6 at the end of this report, we recommend further research with people in that older age group.

As Table 4 shows, education level did correlate with accuracy. Participants with less education made more errors and this result is statistically significant (p<.004).

Table 4. Less education correlated with fewer correct votes.

Highest education level achieved	Mean number of correct votes		
	Ballot A	**Ballot B**	**Both ballots**
Less than high school (n = 9)	14.0	14.9	14.4
High school graduate or GED (n = 15)	15.2	15.9	15.6
Some college or associates degree (n = 12)	15.4	16.7	16.0
Bachelor degree (n = 8)	17.6	17.1	17.4
Some courses beyond college (n = 1)	18.0	16.0	17.0

GED = General Education Development, a series of tests that people can take to show they have the equivalent of a high school education. Many people who drop out of high school take the GED later in life.

The one participant with the highest education level voted Ballot B first. That this participant got higher accuracy on Ballot A as the second ballot suggests that she learned from her experience on Ballot B, as others did.

Education made a slightly greater difference for Ballot A than for Ballot B

Education level was significantly correlated with the number of errors participants made (r = -.419, p < .004, effect size R^2 = 0.176). Less education was associated with more errors. This correlation was slightly stronger with Ballot A – traditional language (r = -.393, p < .008, R^2 = 0.154) than it was with Ballot B – plain language (r = -.359, p < .015, R^2 = 0.129. An analysis of variance (ANOVA) revealed that the difference between the impact of education on accuracy for Ballot A and the impact of education on accuracy for Ballot B, while a trend, was not statistically significant ($F_{4,40}$ = 1.114, p < .364).

Participants recognized the difference in language

The answer to our second question, "Do voters recognize the difference in language between the two ballots?" is also "Yes."

As we show in the next section, the participants overwhelmingly preferred Ballot B overall and for most of the individual pages. The only difference between the two ballots was the language. Therefore, it is fair to assume that the major motivator for their preference was the difference in language.

We also recorded comments from the participants as they voted the two ballots and as they compared the pages. We heard spontaneous comments about the difference in language while participants were voting. In the interview after they voted, we asked participants to focus on the words on the pages. But we did not tell them what to notice or how to voice their comments. Using the script, the moderator only said:

> Notice that the instructions on these pages are different. Please compare them and comment on them.

The fact that participants' reasons for their choices were almost always in terms of plain language guidelines indicates that they were reacting to the aspects of the language that we hypothesized would make a difference.

Here are just a few examples from three pages of the two ballots showing how participants characterized the differences in the instructions:

Comparing the instructions to voters (first screen of each ballot):

A3
On A: I don't like the paragraph being so large and all together.
On B: I like the bullets and that the important points are in bold.

A6
On A: The paragraph form is so long. I gotta read all of this. (with big sigh)
On B: I prefer this; it's less wordy.

B17
On A: When I first read this, I was overwhelmed. I had to read it three times. There was so much to remember.

Comparing the pages about State Supreme Court Chief Justice where A uses "Retention Question" and "retain" and B names the office and uses "keep":

A4   "Keep" is short and sweet compared to "retain." Some people might not know what that ["retain"] means.

B28   This ["Retention Question"] is a little confusing.

C32   "To keep." Yes, yes, I do [want to keep her]. Like I'm thinking 30 seconds less.

Comparing "accept/reject" to "for/against" as choices for measures:

B15   I prefer "for/against"; they are simpler words.

B23:   I prefer "for/against"; it's what a normal voter would say; it's a more commoners' level.

C35:   "For/against" are more common words than "accept/reject."

Participants overwhelmingly preferred the plain language instructions

Both in the page-by-page comparison and in their final, overall assessment, the participants chose the plain language ballot most of the time.

In the page-by-page comparison, participants preferred the page from Ballot B most of the time

On 12 of the 16 pages in the comparison, participants selected the Ballot B page more than 60% of the time. For those pages, the participants' choice of B ranged from 64% to 98%.

On 4 of the 16 pages in the comparison, the participants' choice was very close between the two ballots – and on 3 of those 4 pages, Ballot A was preferred more often (ranging from 51% to 56% of the participants).

We look in more detail at what participants had to say about the language on specific ballot pages in Part 4, Discussion: Which ballot did participants prefer in a page-by-page comparison?

A large majority (82%) of participants chose Ballot B for their overall preference

The answer to our third question, "Do voters prefer one ballot over the other?" is a resounding "Yes" in favor of Ballot B, the ballot with plain language instructions.

82% (37 of 45 participants) chose Ballot B for their overall preference. Just 9% (4 of 45) chose Ballot A, and 9% (4 of 45) chose "no preference." The choice of the plain language instructions for ballots is statistically significant (p<.001).

Participants gave us this preference in writing. The short questionnaire asked them to select one of the three choices (A, B, No preference) and to write down the reasons for their choice.

Ballot B

The 37 participants who chose Ballot B ran the gamut from voting perfect ballots on both A and B to the person who made the most errors across both ballots. These quotes from their questionnaires are typical of the reasons participants gave for preferring Ballot B:

A2:   It helped me understand better so I would not need help.

A5:   Ballot B was more clear and precise. It had better instructions and was easier to follow.

B25:   Ballot B is more specific.

B27:   Ballot A is not very easy to use and wastes more time because it is slightly confusing. Ballot B is not too confusing.

C43:   Easier for me to understand.

C45:   B seems easier to read and follow the instructions.

Ballot A

The four participants who chose Ballot A included two who voted perfect ballots both times and two who made a medium number of errors (5 and 7 errors respectively). These participants seemed to focus on the fact that Ballot A had fewer words.

A6:   Ballot A had less for me to read and seemed easier to understand.

B26:   Ballot A is more easily understandable.

No preference

The four participants who chose no preference included one who voted a perfect ballot both times, two who had just two errors across the two ballots, and one who made 12 errors. In general, they were people who found specific things they liked on each ballot.

A7: Many similarities. And there were minor things in both that I liked better than the other and minor dislikes about both.

A13: There are a few changes that need to be made in both A and B.

Although Ballot B was better than Ballot A, it was not perfect

Although participants did much better on Ballot B than on Ballot A, they were not entirely successful with either ballot. We explore the errors on both ballots and how language may have affected them in Part 3, Discussion: Where did participants have problems?

On a few specific pages, more participants preferred the Ballot A version. We can ask: Are those results a statement against plain language? The answer: "No."

There is a tension between putting as few words on the screen as possible and being specific enough to be useful. On many pages, the Ballot B instructions included more detail than the Ballot A instructions (following the guideline to "be specific; give people the information they need"). Many participants commented on many of those pages that the more specific details were useful.

A clear example is the final page of the ballots. Ballot A just said "Thank You." Ballot B said "Thank You" and then had two separate sentences after that: "Your vote has been recorded. Thank you for voting." All but one participant (44 of 45) preferred the Ballot B page because of the clear, plain language sentence that gave them what they needed: reassurance that they had successfully completed the task.

In one case, we had more detail on the Ballot A version. The page for the President/Vice President contest on Ballot B had just the instruction "Vote for one." On Ballot A, it had the same instruction plus the extra sentence: "(A vote for the candidates will actually be a vote for their electors.)" We put the extra sentence on A in this case because we thought it was superfluous information that some people might not understand. They might not know the word "electors" or might find the concept difficult. In the comparison interview, more participants (25 of 45) chose the A version, citing the extra information as being useful. But almost as many (20 of 45) chose the B version, saying that the extra information was not needed and either confused them or might confuse others.

In Part 4, where we give a page-by-page description of the differences between the ballots and what participants said about both versions, you will see that the pages with the greatest difference in traditional compared to plain language received the highest percentage preferences for the plain language version. The pages that were close in preference had relatively little difference between the two versions.

In Part 5, Recommendations, we list changes that we would make to Ballot B to produce an even better plain language ballot than the one we had in this study.

Part 3
Discussion: Where did participants have problems?

In this section, we look in detail at what happened when our participants voted on the two ballots. We are interested in these two questions:

- Did the language of the instructions help or hinder participants as they voted?

- What besides the language of the instructions was helping or hindering participants?

As we look in detail at what happened, we make use of both

- data on recorded errors (the performance that is counted for the results in Part 2 of this report)

- observations from our notes and from our video/audio recording of the participants as they worked with the ballots

Other studies looked at error rates after ballots were cast

If we are interested in how usable ballots are for voters, one measure of usability is effectiveness, and counting errors is one way to look at effectiveness.

Most research studies about voting look at residual votes (undervotes and overvotes) as errors. However, those researchers are reviewing ballots after an election. They rarely know why the errors happened. Did voters simply choose not to vote in a particular contest? Did they not understand the instructions on the ballot or in the help? Was the design hindering them? What specifically about the language or design was a problem? Research that focuses on already-cast ballots can only speculate.

Error rates are tantalizing evidence that something is wrong. Sometimes specialists in design, language, or usability can make strong educated guesses about what went wrong.[9] In most post-election studies, however, we do not know what effect any specific aspects of ballot language or design had on voters' behavior.

---

9 For example, we have a very good idea about what probably caused the problems with the "butterfly ballot" in 2000 and with the unusual undervote in a Sarasota County, Florida election in 2006 where two contests appeared on one screen. See Norden, Lawrence, David Kimball, Whitney Quesenbery, and Margaret Chen, *Better Ballots*, Brennan Center for Justice at NYU School of Law, 2008. Available at
http://www.brennancenter.org/content/resource/better_ballots/

We observed voting, giving us a better understanding of why voters make mistakes

In our study, we were able to observe people as they voted. Just by observing the act of voting, we learned a lot about when and how our participants had trouble with these ballots. In addition, many participants talked as they were voting about what they were doing and why they did what they did.

In this part, we use mistakes that participants recovered from as well as "official" errors

In this study, an *error* is casting a vote in a way that is contrary to the directions we gave participants. These are the errors that make up the accuracy (performance) data that we reported in Part 2: Results.

In addition, we observed voters making mistakes as they were voting and then recovering from those mistakes. Problems from which our participants recovered are not seen in the error data that we used for our statistics. However, they are informative.

Here in Part 3, when we give statistics on *error rates*, we are using the same data that we reported in Part 2. When we give numbers that include participants whose behavior was interesting and informative even if, in the end, they succeeded in voting correctly, we will be clear that we are including more participants than are seen in the official error count for that contest.

How we organized Part 3

We begin Part 3 with some more details on

- how participants performed (numbers of participants by number of errors on a ballot)

- how pages performed (number of errors on specific pages of the ballots)

We find that six ballot pages account for an overwhelming number of the errors, and we then consider four general reasons why those pages have the highest error rates.

The rest of Part 3 is a detailed discussion of those six pages plus two more ballot pages. The additional two ballot pages are

- the second straight-party voting page (on which participants decided whether to review or bypass/skip party-based races)

- the Summary/Review page at the end of the ballot

Those two pages do not show up in the error count at all because the actions on those pages are only navigational and not actual votes. But we saw participants struggle on both of those pages. We combine the discussion of the second straight-party voting page with our discussion of straight-party voting. We give the Summary/Review page its own section at the end of Part 3.

How participants performed

Table 5 shows how many participants for each number of errors for each of the ballots.

Table 5. How many participants made how many errors.

Number of errors on the ballot	Number of participants making that number of errors	
	Ballot A	Ballot B
0	13	12
1	9	14
2	5	5
3	7	7
4	3	1
5	2	3
6	2	1
7	1	2
8	1	
9		
10	1	
11		
12		
13	1	

Participants across the spectrum voted perfect ballots

Six participants made no errors on either ballot. Seven more made no errors on A but one or more errors on B. Six more made no errors on B but one or more errors on A.

When we look at the participants' education, age, and voting experience, we see that the spread, particularly of education, is greater for perfect ballots on Ballot B than on

Ballot A. Perfect ballots on A tend to cluster at the higher education levels. This may indicate that people with more education can do well on a ballot with traditional language but that ballot does not support people with less education, while people with a wider range of education can do well on the plain language ballot.

About participants who made no errors on <u>both</u> ballots (n = 6)

- 2 had not finished high school (and were younger (18-25) and had not voted before)
- 1 had a high school degree (and was older (51-60) and had voted several times before)
- 1 had some college (and was young (26-30) and had voted a couple of times before)
- 2 had college degrees (and ranged in age from 26-50 and had voted a few times before)

About participants who made no errors on Ballot A (n = 13, including the 6 who voted perfectly on both ballots)

- 2 had not finished high school (and were younger (18-25) and had not voted before)
- 2 had a high school degree (and were older (51-60) and had voted many times before)
- 3 had some college (and ranged in age from 18 to 50 and had voted a couple of times before)
- 5 had college degrees (and ranged in age from 22 to 50 and had voted before)
- 1 had some course beyond college (and was over 60 and had voted many times)

About participants who made no errors on Ballot B (n = 12, including the 6 who voted perfectly on both ballots)

- 3 had not finished high school (and were younger (18-25) and had not voted before
- 3 had high school degrees (and ranged in age from 31 to 60 and had voted a few times before)
- 4 had some college (and ranged in age from 18 to 40 and voted a couple of times)
- 2 had college degrees (and ranged in age from 26-50 and had voted a few times before)

Most participants made relatively few errors

As Table 5 shows, most participants made four or fewer errors on either ballot. Table 5 also shows that the number of errors on B clustered at the lower end compared to A, and three participants made more errors on A than anyone made on B.

A few participants are responsible for a sizable portion of the total error count

Five participants made 35.7% of the errors (70 of 196 errors).

Of those, three did not complete high school; two completed high school.

Adding five more people to the five above, we find that 10 participants (that is, 22% of our participants) made more than half of the errors (55.6%; 109 of 196 errors).

Of those, five did not complete high school; four completed high school; one completed some college but had never voted before.

Once again, we find that education level was negatively correlated with errors.

How pages performed: Six pages had very high error rates

On every page, at least one participant made an error, but it is the pages with the highest error rates that concern us most.

Table 6 shows the six pages that had an error rate of 13.3% or higher for both Ballot A and Ballot B. The other 12 pages had error rates of 6.7% or less.

Table 6. The six pages with the highest error rates

	Error Rate	
Page	**Ballot A**	**Ballot B**
Straight Party Vote/Straight Party Voting	22.2%	15.6%
US Senate	20.0%	15.6%
Registrar of Deeds (where participants were to write in a candidate)	28.9%	26.7%
State Senator (where participants were to change a vote)	26.7%	26.7%
County Commissioners	22.2%	13.3%
City Council (where participants were to change a vote)	26.7%	28.9%

Why did these six pages have such high error rates?

The tasks that were involved in these pages were:

- Vote straight-party and then review the individual races to be able to change some.

- Write in a candidate for Registrar of Deeds.

- Change from the straight-party candidate selected for State Senator to a candidate from a different party.

- Change at least one of the male straight-party candidates selected for City Council to a female candidate from another party.

- Leave the straight-party candidates for US Senate and for County Commissioners. (Note that this was not explicitly stated in the directions we gave participants on how to vote. It was implicit in that we did not tell them to make any changes in those contests.)

Several reasons almost certainly contribute to the high error rates on these pages, including these four:

- Voters must know a lot about how elections work to follow a ballot.
- Experience with other technology does not necessarily carry over to give voters a good mental model of using an electronic voting system.
- Many voters do not understand different levels of offices.
- Electronic voting systems compound the problem because voters do not know what is yet to come on the ballot.

Voters must know a lot about how elections work to follow a ballot

For voters to successfully cast a ballot with the votes they want, they must come to the ballot with knowledge about political offices, elections, voting, ballots, and voting systems.

- Some voting tasks require voters to understand concepts that are specific to elections, such as voting a straight-party ticket, writing in candidates, and choosing fewer than the maximum number of candidates in multi-candidate contests.
- Ballots assume that voters understand the different levels of contests (from federal to local) and the difference between partisan and non-partisan contests – whether the ballot uses the words "partisan"/"non-partisan" or not.

Experience with other technology does not necessarily give voters a good mental model of using an electronic voting system

Voting systems do not operate in the same way as other technology that voters are familiar with.

Experience with options in software does not necessarily help. Electronic voting systems don't operate like radio buttons or checkboxes in software. In an electronic voting system like the one we simulated, you cannot change your choice just by clicking on a new choice (as you would in a radio-button interface). If a choice is already selected, you must click on the *selected choice* to "deselect" it before the system will register a different choice.

Experience with other touch-screen systems like ATMs and travel kiosks does not necessarily help. Again, the interaction design is not the same. Most of our participants (37 of 45) indicated that they regularly use ATMs. Although they may have known from that experience how to use a touch-screen, the mental model of the process of getting money from an ATM is not the same as the mental model needed to understand voting. Using an ATM is often a single process. Voting is a series of individual contests, each of which may have different rules (such as how many people you may vote for).

Moreover, the infrequency of voting and the potentially significant consequences of "getting it wrong" may raise voters' anxiety level in ways that other encounters with technology do not.

Many voters do not understand different levels of offices

A single ballot may include (as ours did) contests on the federal, state, county, and city levels. And many ballots include (as ours did) contests that are party-based and that are not party-based.

Many of our participants had difficulty distinguishing between the contests for US Senate and State Senator and between the contests for County Commissioners and City Council. (See the sections below on those contests for more details.)

We do not know if this is because schools are not teaching civics, because our participants didn't take classes in civics, because they took those classes a long time ago, or for some other reason. We do not know if it is because they have not received voter education materials in past elections, because they did not pay attention to those materials, or for some other reason. We do not know if it is because even though most said they had voted before, they have only voted for the federal offices and have generally ignored offices on other levels of government, or for some other reason.

Electronic voting systems compound the problem because voters do not know what is yet to come on the ballot

The progressive disclosure of the typical electronic ballot doesn't help voters. On an electronic ballot, voters see only one contest at a time and have no information about the contests that come later on the ballot. The US Senate contest that many participants mistook for the State Senate contest comes earlier in the ballot. Similarly, the County Commissioner contest that many participants mistook for the City Council contest comes earlier in the ballot.

Furthermore, the cognitive load of using the system may have been enough of a drain on attention to prevent some of our participants from returning to contests earlier in the ballot to correct errors as they realized them later.

Did plain language make a difference?

As we explain in more detail in the following sections, for some of these six pages, plain language made a positive difference. This was especially true at the beginning and end of the ballot. "How to Vote" on the first main page of Ballot B, the instructions for voting straight-party, and the instructions on the Review Your Choices page all helped participants more than the instructions on the comparable pages of Ballot A did.

However, based on our observations, we could improve the plain language of our Ballot B in ways that would make those instructions work even better for voters. We describe those changes in Part 5, Recommendations for creating a ballot that voters can understand and use successfully.

For a few of the six pages with the highest error rates, language was not a major factor in the problems that our participants had. For example, the language on the US Senate contest and the State Senator contest were identical to each other and identical on both ballots: Vote for one. It was not the language that made people mistake one office for another.

Pages   **Straight Party Vote/Straight Party Voting**

There are two pages for straight-party voting on the ballots that we tested. The first page presents the party choices with instructions for voting straight-party. The second page confirms the party choice and gives instructions and options for either accepting all the party-based votes now selected automatically or reviewing each party-based contest with the option of changing votes along the way.

The directions told participants quite explicitly to select a party (Tan on A, Lime on B) on the first straight-party voting page:

Task   Ballot A:
You usually vote for everyone in the Tan party.
Vote for all the people in that party at one time.

Ballot B:
You usually vote for everyone in the Lime party.
Vote for all the people in that party at one time.

Although there was no specific task for the second straight-party voting page, by reviewing the next several tasks on their sheet of directions (which each participant took time to read through before starting the ballot), participants should have realized that they had to look at the party-based contests.

What happened

The error data refers only to the first straight-party voting page.

Errors

Straight Party Vote	Ballot A	22.2%	10 of 45
Straight Party Voting	Ballot B	15.6%	7 of 45

Participants were more likely to correctly select straight-party voting on Ballot B (84.4% correct) than on Ballot A (77.8% correct).

One of the errors on A is a participant who chose straight-party but chose the wrong party. All of the other errors in this data for the first straight-party voting page are people who chose not to vote straight-party. Of our 45 participants, 13 chose not to vote straight-party on one or both ballots: 3 on both; 6 more on A; 4 more on B.

In a real election, that would not actually be an error. Voting contest by contest would be acceptable. We coded it as an error because it was contrary to our directions and was an indication that the language on the ballot was not helping people understand the options for and implications of voting straight-party and then changing party-based contests.

Because the choice that participants made on the second straight-party voting page was only navigation – go to the next contest or skip over many contests, nothing was recorded as a "result" in the computer-based data file. Information on what happened on the second straight-party voting page comes from our observations and notes. These show that of the 33 people who saw that page on A, 9 chose (incorrectly) to bypass the party-based contests. Of the 38 people who saw that page on B, 8 chose (incorrectly) to skip the party-based contests.

What we learned

Straight-party voting is a difficult concept for many voters. Many voters are not familiar with it. Providing a straight-party option on ballots is illegal in some states. New Hampshire outlawed straight-party options while we were working on this report. In most states where straight-party voting is allowed, it is up to each county to decide whether to use that option. (Of course, in all states, voters may give all their votes for party-based contests to people from the same party by voting for that party in each separate contest. When a ballot has no straight-party option, it is only taking away the choice of voting once and having that vote apply automatically to all candidates of that party.)

Many voters may not be aware of which contests below the federal level are party-based and which are not. Some comments from participants suggested that they expected all contests to include party affiliations.

Some of our participants did not know the meaning of "partisan." When ballots use "partisan" and "non-partisan" to distinguish different sets of contests, many voters may be confused because they don't know what the word means.

Being able to change a straight-party choice is a difficult concept for many voters. By tradition, once you have voted straight-party, you're committed; there's no changing. On a paper ballot with straight-party voting, there are only two options: vote straight-party without marking votes in individual party-based contests or vote only in individual contests.

However, electronic voting provides the option for voters to change their votes in individual contests after selecting to vote a straight-party ticket. While this type of interaction is common in software and web applications – clicking one button to "select all" with the option then to "deselect" any one item in the list – the idea does not seem to map readily to electronic ballots.

So, there's a conflict in logic when there is the possibility of voting straight-party and then being able to change some of those votes.

Did plain language help?

Yes, but…

Instructions on Ballot B included explanatory text about what a straight-party ticket is, that you could vote straight-party and change individual party-based contests, and how to do it.

The explanatory text and instructions helped most participants. Participants were much more likely to vote correctly on the straight-party task on Ballot B regardless of the ballot they used first. And they preferred Ballot B by a wide margin. (See Part 4 Discussion: Which pages did participants prefer?)

Some, to be on the safe side, chose not to vote straight-party but instead voted each contest. Participants were much more likely to do this on Ballot A than on Ballot B, especially if they used Ballot A first. This suggests that they could not tell from the instructions on A how to complete what they needed to do – both vote straight-party and also vote for someone in a different party for one of the party-based contests.

However, despite the plain language instructions on Ballot B that helped many participants, we still see a high error rate on Ballot B.

In part, that is probably due to the difficulty of the concept that a voter can vote straight-party and then change individual contests.

In part, that is probably due to the problems we saw people having with the length of the instructions and confusion over the wording of the button choices on the second straight party voting page on Ballot B. We suggest new wording for the second straight party voting page of a plain language ballot in Part 5: Recommendations. **Plain language definitely helped, but the results show that the language on Ballot B could have been even clearer and plainer.**

Pages	**US Senate, State Senator**

| **Tasks** | Ballot A: |

Ballot A:
For State Senator, instead of the Tan party person, you want the Orange party person. Make sure your vote for State Senator is for the Orange party person.

Ballot B:
For State Senator, instead of the Lime party person, you want the Purple party person. Make sure your vote for State Senator is for the Purple party person.

There was no task or direction to change a vote for US Senate. The correct choice for US Senate was to leave the person selected as the straight-party candidate.

What happened

The instructions on the US Senate and State Senator ballot pages for both versions of the ballot were the same: Vote for one.

A sizable portion of participants voted correctly on both ballots for both of the senate contests: 18 participants (40%) voted correctly in all four cases.

However, 27 participants (60%) voted incorrectly on at least one senate contest on at least one ballot. Of those, 15 participants (33.3%) voted wrong for US Senate on at least one ballot.

In addition to the error data we show here, the observational data – behavior of participants and comments they made as they performed the tasks – tell us things that looking at the error data alone would not. On this task several participants asked either themselves or the moderator-as-poll worker whether US Senate was the same as State Senator.

Errors US Senate	Ballot A	20.0%	9 of 45
	Ballot B	15.6%	7of 45
Errors State Senator	Ballot A	26.7%	12 of 45
	Ballot B	26.7%	12 of 45

What we learned

Some participants weren't clear or confident about the difference between the US Senate contest and the State Senator contest. We infer that those who changed their votes on the wrong contest were simply unsure where in the ballot to make the change. (Several participants, upon arriving at the US Senate contest, looked at the directions and asked themselves if this was the race to change. Some decided it was even though the directions said "State Senator.")

Participants were more likely to make this mistake on their first ballot. That supports our belief that the progressive disclosure of the electronic ballot caused part of this problem. Because participants had no way of knowing whether there was another contest that more closely matched the one they were looking for, they performed the task on the first contest that seemed to be right.

Going through one ballot helped some participants understand the hierarchy of the ballot. Lower error rates on the second ballot suggest that participants learned about how the ballot worked overall the first time around, even when they did not correct mistakes on the first ballot.

Participants did not go back to correct mistakes in the earlier contests. It is difficult to know why participants did not return to the incorrectly voted US Senate contest to correct their choices. Based on our observations of behavior and comments we heard, it seems likely that some participants were very focused on moving forward in the ballot. Others said they expected an opportunity to correct mistakes before finally casting their ballots. A few participants did use the Summary/Review page this way, but others seem to have forgotten when they got there that they had an incorrect vote early on in the ballot. Or, it may be that a few participants were so taxed by everything they had to manage during the session that they forgot to go back – the cognitive load factor we mentioned on page 35.

Participants were able to change votes. To vote incorrectly on US Senate or to vote correctly on State Senator required most participants to change a vote – that is, to "deselect" an already selected candidate by clicking on that highlighted name and then to select a new candidate by clicking on that person's name. (Only participants who chose not to vote straight-party and to go through the contests reached these contests with no candidate previously selected.)

The error rates showing how often people changed US Senate incorrectly as well as the non-error rates showing how often people changed State Senator correctly suggest that participants were able to change votes. However, we observed many people having problems deselecting – either because they had forgotten that they needed to do that or because they had problems getting the touch screen to accept their taps. Several participants used the ? [help] option to review the instructions on how to select and were then reminded of the need to deselect. In the end, however, although almost all of our participants had problems both in remembering to deselect and in getting the touch screen to respond, most were successful at changing votes.

For State Senator, some problems may have been a ripple effect from the Registrar for Deeds task. In addition to issues related to level of office, it is likely that some errors on State Senator came from other factors. For example, some issues with writing in on the Registrar of Deeds page rippled to the next page in the ballot, State Senator. In trying to get to a write-in page, some participants clicked Next, returned to Registrar of Deeds, and cycled back though the pages again, becoming disoriented. (We discuss the problems with writing in later in Part 3 in the section on the Registrar of Deeds page.)

The artifact of the test may have contributed to the confusion between the two contests. Although the ACCURATE team had previously shown that study participants were just as accurate in voting a ballot with made up names as a ballot with real names, it is possible that some of the errors we saw in our study come from the artificiality of the names of both parties and candidates. We do not know how well voters in real elections know their preferred candidates for US Senate and State Senator contests. The task in our study told participants which contest to change but they were changing from one unfamiliar party to another and the names of all the candidates for those contests were not known to them beforehand. It would be instructive to do usability tests with real ballots and observations of real voting situations to know if this serious confusion that we saw between the federal level contest (US Senate) and the state level contest (State Senator) is truly an issue in many elections.

Did plain language make a difference?

The instructions on the US Senate and State Senator contests were the same on both ballots: Vote for one. Language itself cannot have made a difference on the four specific pages for those two contests on the two ballots.

However, plain language instructions at the front of the ballot helped participants change their votes.

Participants noticed and read the instructions for changing choices on the opening page of the ballot (Instructions to Voters/How to Vote). Many participants remembered these well enough to change votes later. A few participants used the ?

[help] option or backed their way to the beginning of the ballot to read the instructions again.

In the preference data, you will see that participants greatly preferred the plain language instructions on Ballot B (How to Vote) over the traditional instructions on Ballot A (Instructions to Voters). Participants had some difficulty finding the instruction for changing choices within the long paragraph on Ballot A. When they got to Ballot B after using A, a few participants commented that they had not remembered seeing similar instructions on A.

Our conclusion on the value of plain language for the US Senate and State Senator contests: **Plain language instructions helped voters change choices but could not help them understand the hierarchy of the ballot.**

Page	**Registrar of Deeds**
Tasks	Ballot A: Even though you voted for everyone in the Tan party, for Registrar of Deeds, you want Herbert Liddicoat. Vote for him.
	Ballot B: Even though you voted for everyone in the Lime party, for Registrar of Deeds, you want Albert Sterner. Vote for him.

What happened

More than half of the participants (24, or 53.3%) voted correctly on both ballots for Registrar of Deeds.

Six participants (13.3%) voted incorrectly on both ballots for this contest.

The rest (15, or 33.3%) voted incorrectly on Registrar of Deeds on one ballot or the other.

Participants voting on Ballot A first were more likely to vote incorrectly on Registrar of Deeds than participants voting on Ballot B first.

Errors Registrar of Deeds	Ballot A	28.9%	13 of 45
	Ballot B	26.7%	12 of 45

In addition to the error data, we observed 14 participants having problems on the way to completing the task.

What we learned

Participants realized they had to write in the candidate. When they arrived at the Registrar of Deeds page and saw that the person they wanted to vote for was not a listed candidate, all participants realized that they would need to write in the person's name. The problem here was not related to the idea of writing in a candidate; it was primarily related to how to accomplish that with the DRE voting system.

Deselecting is a difficult concept. Several participants did not remember how to change choices. This was the first task in which participants should have deselected a candidate to change a vote. Some participants faced deselecting earlier when they chose incorrectly to change the vote for US Senate. For those who left US Senate unchanged (the correct choice), Registrar of Deeds was the first page where deselecting became an issue.

Even though they had read the instructions at the beginning of the ballot, most participants attempted to change candidates without first deselecting the already-selected candidate at least once (and in some cases many times). Although most recovered by either re-reading the instructions or remembering on their own, the experience was frustrating to them.

According to our observational data, deselecting was extremely frustrating for 14 participants. (There were repeated attempts to touch the choice without deselecting first.) Data from the video recordings and our notes show that four of those 14 who had serious difficulty recovered on their own. The other 10 participants who had problems did not recover on their own. Many asked for help from the moderator-as-poll worker, went to ? [help] on their own, or gave up.

Some participants reviewed the ballot instructions to learn how to change votes. Six participants, either through prompting from the moderator when they asked the moderator-as-poll worker a question or on their own, reviewed the instructions on the first instructional page in the ballot. (The moderator, serving as "poll worker," responded to questions by asking if there was some place in the ballot where the participant had seen information that might meet the need. This question was sufficient to get participants to click on the ? icon or to back up, and both of those actions took them to the initial page of instructions that they had seen at the beginning of the ballot.) However, in this case, the instructions were not as helpful as they could have been. We discuss this below under the heading, Did plain language make a difference?

The touch-screen interface was problematic for several participants. Even when they knew that they had to deselect a candidate, many of our participants had difficulty getting the touch-screen to register their taps (both for deselecting and for selecting). We understand that this is a problem on actual DREs as it was on our tablets.

Problems in deselecting and selecting caused some participants to seek other ways to accomplish a write in. When they could not get any action by tapping on the box labeled Write-In Candidate/Write in a candidate's name, interestingly, three participants attempted to use the stylus to write directly onto the screen on the Registrar of Deeds page. (Others verbalized whether writing on the screen was the way to do it, but didn't try.) These three participants eventually recovered to be able to write in the correct candidate on Registrar of Deeds on B.

Did plain language make a difference?

Yes, and no.

Using the Registrar of Deeds page successfully for this task required participants to have read and understood the instructions at the beginning of the ballot.

The instruction for writing in a candidate was buried inside the dense paragraph on Ballot A, and that did not help participants. Figure 2 shows the instructions from the beginning of Ballot A.

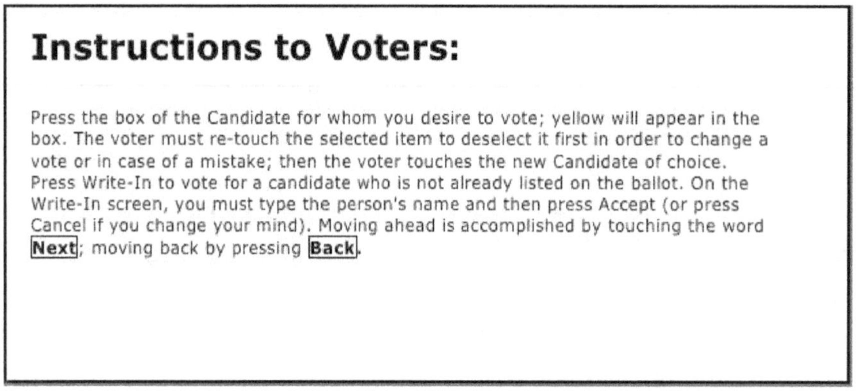

Instructions to Voters:

Press the box of the Candidate for whom you desire to vote; yellow will appear in the box. The voter must re-touch the selected item to deselect it first in order to change a vote or in case of a mistake; then the voter touches the new Candidate of choice. Press Write-In to vote for a candidate who is not already listed on the ballot. On the Write-In screen, you must type the person's name and then press Accept (or press Cancel if you change your mind). Moving ahead is accomplished by touching the word Next; moving back by pressing Back.

Figure 2. The instructions at the beginning of Ballot A.

As the preference data show, participants preferred the plain language instructions of Ballot B. Figure 3 shows the instructions from the beginning of Ballot B.

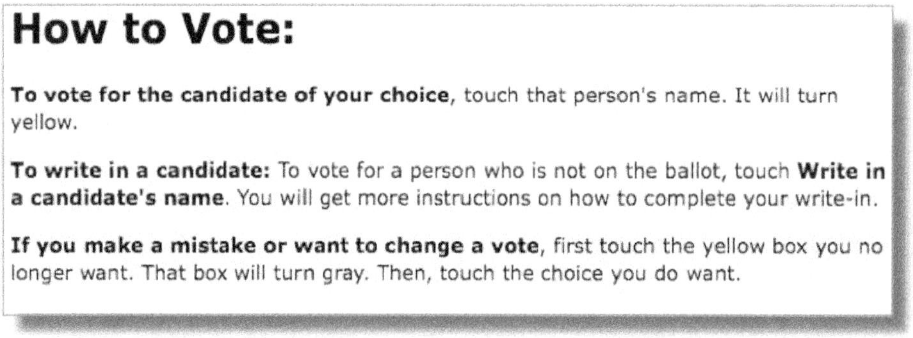

How to Vote:

To vote for the candidate of your choice, touch that person's name. It will turn yellow.

To write in a candidate: To vote for a person who is not on the ballot, touch **Write in a candidate's name**. You will get more instructions on how to complete your write-in.

If you make a mistake or want to change a vote, first touch the yellow box you no longer want. That box will turn gray. Then, touch the choice you do want.

Figure 3. The instructions at the beginning of Ballot B.

Participants wanted the instructions broken into separate sections. They wanted the bold. They liked the title of the Ballot B instructions. However, for the specific needs of the write-in task, the plain language instructions as we had written them also had a problem.

A few participants went back to these instructions several times when they were trying to write in a new candidate for Registrar of Deeds. They were so focused on that task that they only read the middle paragraph. They saw the bold lead-in for writing in, read that paragraph, and stopped. They went back to the contest page, missing the instruction that would have helped them most – the one about deselecting.

While the plain language instructions helped in many ways, for this task they were in the wrong order. We give specific recommendations for improving these plain language instructions in Part 5: Recommendations.

Page   **County Commissioners**

Tasks

There was no task or direction to change a vote for County Commissioners.
The correct choice for County Commissioners was to leave the people selected
as the straight-party candidates.

What happened

Errors County Commissioner   Ballot A   22.2%   10 of 45

Ballot B   13.3%   6 of 45

What we learned

Again, participants seemed unclear about the different levels of government.
The County Commissioner and City Council contests were next to each other in the
ballot in that order. Both were party-based contests that allowed participants to vote
for more than one candidate. If participants voted straight-party, both contests had
candidates already selected when participants arrived at those pages of the ballot.

Most of the errors on County Commissioners were due to participants mistaking that
contest for the City Council race. Participants changed votes on the County
Commissioner contest to include more women – the task that we had given them for
the City Council contest. We observed and heard several participants ask
themselves or the moderator-as-poll worker if County Commissioners was the same
as City Council. (The moderator-as-poll worker did not answer the question. She
asked the participant what he or she thought.) This is another instance of the same
problem we saw and heard between US Senate and State Senator: participants
being unsure of different contests and not being able to look ahead in an electronic
ballot.

**A few participants had questions about how many candidates they could vote
for.** The instruction for County Commissioners on Ballot A was Vote for no more than
five. The instruction on Ballot B was Vote for one, two, three, four, or five. One
participant asked on Ballot A, "Does that mean you can vote for less than five?" The
same participant asked on Ballot B, "Does this mean that you should vote for
[number]1, [number] 2, [number] 3, [number] 4, or [number] 5 down the list, ignoring
any other candidates?" A few other participants verbalized similar questions either to
themselves or to the moderator-as-poll worker.

Some participants felt the need to vote for the maximum number of candidates. County Commissioners was a multi-candidate contest where voters could select up to five candidates. On each ballot, only three of the candidates were from the party our participants were supporting (Tan on A, Lime on B). If participants followed our directions, they would not change anything on the County Commissioner page. We were implicitly telling our participants to purposefully undervote in this contest. For County Commissioners, changing anything, including adding more candidates, was an error in our study.

Nine participants voted for more than three candidates on either Ballot A or Ballot B or both, and 7 of those 9 used Ballot A first.

Five participants on A and one participant on B added more candidates when they first got to the County Commissioners page. Either they were following directions and were confusing County Commissioners for City Council or they were reacting to the maximum number from the instruction on that page.

Three participants on Ballot A and three participants on Ballot B added candidates by going back to the County Commissioners page from the Summary/Review page. This probably relates to participants feeling the absolute need to not have any remaining red blocks on the Summary/Review page. We discuss this issue later in Part 3 in the section about the Summary/Review page.

Did plain language make a difference?

Most of the errors that participants made on County Commissioner were related to a lack of understanding of the hierarchy of the ballot – to mistaking this contest for the City Council contest that came next in the ballot. This is not a language issue.

A few errors came from participants insisting on voting for the maximum number. As we said above, a few participants said that they thought the instruction on Ballot A meant that they had to vote for five candidates. However, the more common reason for voting for the maximum was to get rid of the red block on the Summary/Review page rather than because of the instruction on the contest page.

Page	City Council
Task	Ballot A and Ballot B: For City Council, you think that the women running are the best candidates, so vote for them. You decide what to do about the other candidates for City Council and tell me what you are doing when you decide.

Errors City Council	Ballot A	26.7%	12 of 45
	Ballot B	28.9%	13 of 45

Eight participants voted incorrectly on both ballots for City Council.

What happened

City Council was another multi-candidate contest – this time with a maximum of four votes. If participants had successfully voted for a party (Tan on A, Lime on B), all four possible votes were taken by their party's candidates. On each ballot, three of those party candidates had masculine names, one had a feminine name. On each ballot, one candidate from another party had a feminine name. (The names were Carol on Ballot A, Carole on Ballot B, and Barbara on Ballots A and B.) To successfully complete the task for this contest, participants had to deselect at least one male candidate before they could select the female candidate from another party.

Because we then allowed participants to do what they wanted for the rest of this contest, we accepted as correct any combination as long as it included the two female candidates. Options participants chose that we counted as correct included: voting for only the two female candidates, voting for the two female candidates and any one or two other candidates, voting for the two female candidates and any one or two write-in votes. (Several women wrote in other women to fill out their four possible votes.)

What we learned

Problems in deselecting and in getting the touch-screen to respond prevented some participants from completing the task. Participants understood the first part of the task – to vote for the two women on the ballot. Although some hesitated about the second half of the task – where we gave them leeway to decide what else to do, their hesitation didn't affect error rates. We accepted any other votes they chose, as long as they included both female candidates.

The incorrect votes were from people who left the straight-party selections (three men and one woman) or who changed votes but only ended up with one woman (not two). Both of these were primarily caused by the same problem we had seen earlier in the ballot – not remembering to deselect or not succeeding at deselecting (or at selecting someone new) because their tapping didn't get a response from the system. Participants who became overly frustrated stopped trying.

A note: The issues we saw in doing the task successfully have counterparts on paper ballots. As we have said, deselecting an already-selected candidate and selecting an unselected candidate were difficult for many participants. They were not sure where or how to touch (tap) the screen to get it to respond. They tried (often several times in several places) before the system registered what they were doing. And, as we have said, we understand that voters sometimes have similar problems on existing electronic voting machines.

Although paper is different, similar frustrations occur with paper ballots.

If a voter wants to "deselect" an already-selected candidate on a paper ballot, the voter has to understand how to do it (request a new ballot), be willing to admit the problem, find an appropriate person, get a new ballot, and start all over again. How many people give up and stay with the selected candidates even when they would prefer to change? That would be exactly comparable to our participants who ended up staying with the three men and one woman when they knew they were supposed to vote for two women.

On the touch-screen, problems with getting the system to record a selection were physical (not touching or tapping hard enough or in the right place). On paper, how many voters do not fill the oval completely or darkly enough, do not match the arrows up correctly, or do not punch through the card deeply enough and go on to the next contest assuming that their vote was fine? Those voters would be exactly comparable to our participants who assumed they had selected a candidate because they had tapped the name and who went on to the next contest not noticing that the name did not become highlighted.

A few people added more candidates from the Summary/Review page.
Participants who ended up with fewer than four candidates selected when they left the City Council page saw that contest in a red block on the Summary/Review page at the end of the ballot. Although their two or three candidates met our criteria for the

contest (two women), the Summary/Review page indicated that the contest was undervoted. As we explain in detail later in Part 3, undervoted contests shown in red were simply unacceptable to some participants. They personally decided that they had to go back and fix those contests, even to the extent in one case of recording blank write-in votes. (See the section on Summary/Review page.)

Did plain language make a difference?

The issues on the City Council page, as on the County Commissioners page, were not primarily related to language.

To the extent that language was relevant, the issues were the same as those on other pages: primarily how well each ballot helped participants understand how to change a vote. Participants overwhelmingly told us that the instructions for doing that were easier to understand on B. However, the instructions were only at the beginning on both ballots and participants had difficulty remembering the need for the unusual action of having to deselect a candidate when the maximum number has already been reached.

In Part 5, Recommendations, we recommend putting the instruction about the need to deselect on every page of an electronic ballot. Once again, plain language helped, but after this study, we can suggest an even better implementation of the plain language guidelines.

Page	Ballot Summary/Review Your Choices

Tasks

When participants came to this ballot page, we allowed them time to do whatever review they wanted. When they were about to move on (cast ballot), the moderator interrupted them with an additional task:

> Ballots A and B
> (Note that we did not specify who to vote for or how many people to vote for in this contest):
>
> You decide that you should vote for the Water Commissioners, so do that now.

When they had finished that additional task, we again gave them time to do whatever review they wanted. When they were ready to move on, the moderator interrupted them a final time with these tasks:

> Ballot A:
> You realize that you actually wanted Andrea Solis to be your State Assembly person. Change your vote for State Assembly to Andrea Solis.
>
> When you are ready, finish voting as you really would in a real election.
>
> Ballot B:
> You realize that you actually wanted Edward Shipett to be your State Assembly person. Change your vote for State Assembly to Edward Shippett.
>
> When you are ready, finish voting as you really would in a real election.

With that final instruction, we were interested in how they would end the ballot. If they wanted to, they could review their choices again or they could go directly to casting their ballot.

What the pages looked like and why

On many DREs, after participants have voted for all the contests and measures, they see a page that summarizes all the choices they made. Our simulation included a similar page. On Ballot A, the page was called Ballot Summary. On Ballot B, it was called Review Your Choices. We refer to it here as the Summary/Review page.

Our Summary/Review page mimicked the interaction design and user interface elements of many DREs, including:

- Red messages to show undervoted contests and measures
- Blue blocks to show completely voted contests and measures
- Navigation to return to contests and measures to change votes
- Navigation to move forward and cast the ballot

The behavior of the system was the same for both ballots. The instructions were different on the two pages, as shown in Figure 4.

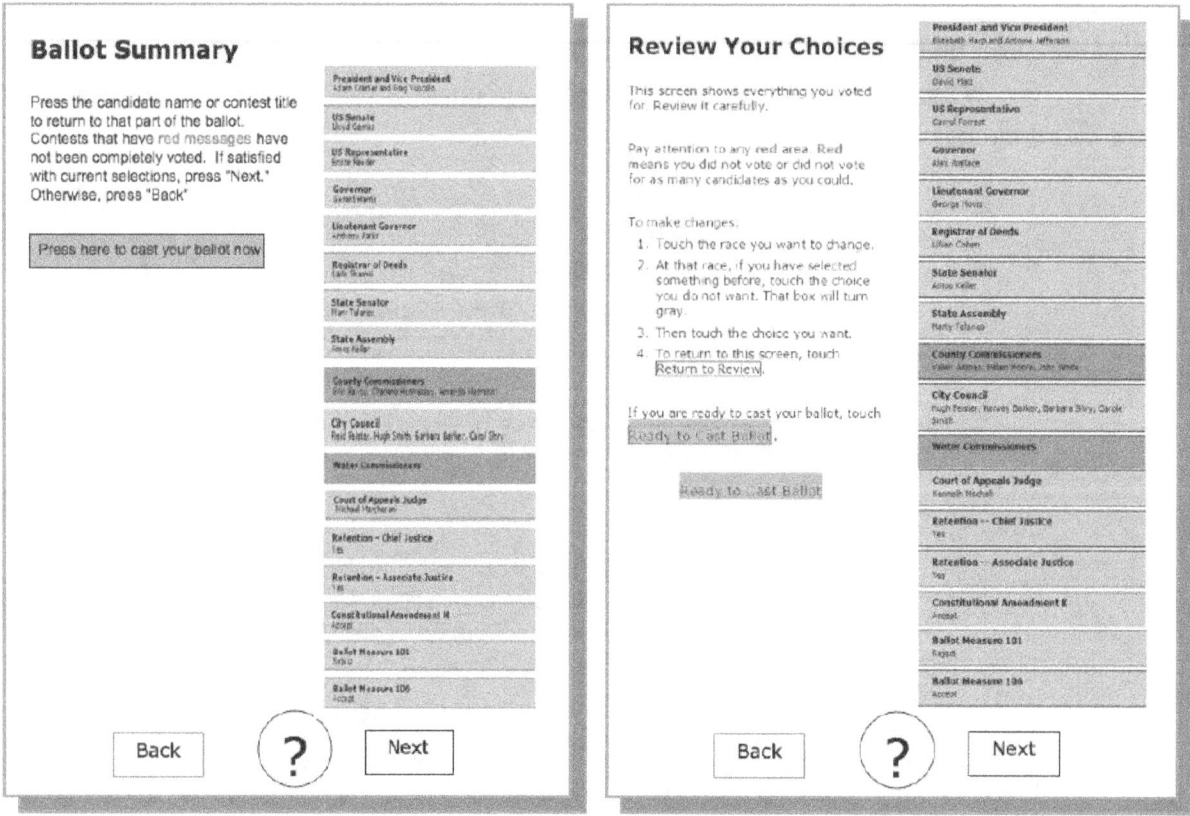

Figure 4. The two Summary/Review pages. Ballot A is on the left. Ballot B is on the right. The user interface design and interaction were the same on both ballots.

We are reporting observational data for these pages

The actions on the Summary/Review page are only navigational and not actual votes, so there is nothing to record from that page in the error count. This is a case where our being able to observe participants helped us see where voters had issues, and we saw participants struggle with many aspects of this page. Therefore, the discussion in this section is based on reviewing our observational data – notes, video recordings, and audio recordings.

What happened

About half of the participants had no problems. From our notes and reviews of the video recordings, 22 participants (49%) had no questions or problems on the Summary/Review page. They were able to reach the end of the ballot having marked the choices as they intended and were ready to cast their ballots.

Of those who had no observable questions or problems, 7 voted on Ballot A first; 15 voted on Ballot B first. This suggests that the instructions on Ballot B were more helpful to participants than the instructions on Ballot A were.

Just over half of all participants had questions or problems. More than half (23 or 51%) did have questions or problems on the Summary/Review page. This is a disturbing number.

Participants were more likely to have questions and problems on Ballot A, regardless of the order in which they used the ballots. Almost twice as many people had questions or problems when voting Ballot A than when voting Ballot B.

- 8 participants had questions or problems on both A and B.
  Of those 8, 6 voted on A first, 2 voted on B first.
- 20 (8 plus an additional 12) had questions or problems on A.
  Of those, 14 voted on A first, 6 voted on B first.
- 11 (8 plus an additional 3) had questions or problems on B.
  Of those, 8 voted on A first, 3 voted on B first.

What we learned

Most of the questions and problems related to the red blocks. Issues on the Summary/Review page were overwhelmingly related to resolving votes shown in red blocks. Although the lack of responsiveness of the screen to participants' tapping was still a frustrating issue to some even on this late page, that only added to the stress caused by trying to deal with the contests that were shown in red.

On the Summary/Review page, undervoted contests and measures were displayed in red blocks. These red blocks were very disturbing to participants.

The directions that we gave participants should have left two contests undervoted by the time they reached the Summary/Review page: County Commissioners and Water Commissioners. That is what you see in the pages in Figure 4 above.

But participants had often undervoted other contests as well. For example, participants who were unsuccessful at getting the system to register their tapping sometimes gave up after a few tries and moved forward in the ballot, leaving a contest unvoted. Participants who had changed the City Council contest to just two female candidates (a correct vote by our directions) saw that contest in red on the Summary/Review page.

Marking undervoted races in red caused some participants to believe they must select the maximum number of candidates. Observational data tells us that 17 participants (37.8%) verbalized questions or concerns about the sections that were shown in red. Reading the instructions cleared up the question for some participants, but the importance of the outcome magnified the need to get the answer right. These quotes from participants demonstrate the issue:

B26: So, what's the problem?

   [Reads the instruction about red messages.] But I did. I did what it told me to do. Red messages are saying that I'm not doing something right.

   I voted for the number of candidates. I'm concerned that it should have turned to blue. That would make me sure that I did the right thing.

   I wouldn't vote because it's telling me I'm not doing the right thing.

C43: Did I do something wrong? I think I did those [contests] right so I'm going to cast my ballot. It wouldn't let me vote for 1, but it took 2. It's not red anymore, so it must be right.

A few participants cast their ballots saying that in a real election they might have asked for more help from a poll worker or abandoned their ballots rather than cast them because they were not sure the votes would be counted as they intended.

Resolving the red messages conflicted with the instructions on multi-candidate races, causing some participants to vote for candidates they didn't want or to work around in other ways. B26 agonized – like others – but didn't add votes to clear the red messages on the Summary/Review page. After reading and re-reading instructions, she determined that having some contests "not completely voted" was acceptable. Though she cast her ballot in the study, she said she might not in a real election.

Several other participants decided to go back to undervoted races to choose or write in candidates for the remaining positions, as one participant said, "to be safe." On County Commissioner, participants could vote for up to five candidates but were directed to vote for only three. For Water Commissioner, participants could vote for up to two candidates, but were directed to make their own decision about how many to vote for. These participants were so unsure that their votes would count in the contests that were undervoted, that they decided to vote for people they otherwise wouldn't have just to make their priority votes count.

Participants' solutions sometimes involved writing in unqualified candidates.
In several cases, participants wrote in unqualified candidates for the remaining positions. It is unclear whether they realized that this would work the way they wanted – that is, their votes for the "real" candidates would be counted but the write-ins would be meaningless.

B29 took a novel approach: She returned to undervoted races from the Summary/Review page to enter *blank* write-ins for the remaining choices. She knew that doing this would not void her votes. Entering the blank write-ins satisfied the voting system's edits so that her entire Summary/Review turned blue. She cast her ballot confident that she had voted as she intended and that her ballot would be counted accordingly.

We hadn't expected participants to write in candidates in contests other than Registrar of Deeds, but they did. There seemed to be two things going on with write-ins for our participants:

- Once they learned how to do it, it became a voting tool.
- Voting seems to be similar to filling out other types of forms: people are compelled to fill in every blank.

There are two hidden "rules" that participants' were addressing by entering write-ins:

- The ballot page must be filled in completely. For example, on the County Commissioners page, they saw that they could vote for five but only wanted three of the candidates on the ballot, so they filled in the write-in blanks for their remaining two choices with friends' or celebrities' names.
- "Red" means an "error" that must be resolved. Writing in (even a blank write in) made the block turn blue, which some participants took as a necessary signal that they had voter properly.

Perhaps we are becoming more conditioned to filling out forms and to doing so error-free. That conditioning conflicts with the purposeful flexibility of voting systems in which it is acceptable to intentionally not vote or vote for fewer than the maximum allowed. While the way many of our participants resolved the problem was harmless to an election – their write-ins would not count but would also not negate their other votes – the belief that the contests must be completely voted (whatever the error) caused participants frustration that could be avoided through better user interface design (not using red) and voter education as well as clear, plain language.

A note about voter education and writing in a candidate: Several participants indicated they thought they could write in anyone, even if that person was not an officially "declared" or "qualified" write-in candidate. This may be true in some places; in most jurisdictions, however, write-in candidates must officially declare themselves, usually by filing some kind of paperwork with the local elections department. Our participants did not know that.

Visual complexity in the page design may have been magnified by the importance of the content. The Summary/Review page drew more spontaneous comments than other pages of the ballot. We asked our participants to talk out loud while they were working, but most were rather quiet as they went through the ballot until the Summary/Review page. Then they talked.

This may not be surprising. In many ways, the Summary/Review page is the most important page in the ballot. There's a lot going on visually. It looks completely different from the other pages in the ballot. It signifies a final commitment that cannot be undone or changed later. So, while selecting choices on contests and measures carries some stress, the cognitive and emotional load increases dramatically on reviewing this final major page.

Showing the undervoted contests in solid red blocks got participants' attention. But many overreacted. They took the red to mean that undervoting was incorrect and dangerous. They feared that their votes would not be counted at all if they did not make the blocks turn blue.

Red was actually sending multiple messages. As is typical in many DREs, our ballots used red to indicate any contest or measure that was not as completely voted as possible. Red appeared whether the participant had made some choices or no choice. Participants could only tell if the message was "you could have voted for more" or "you didn't vote for this at all" by looking carefully at the words in each red block (and text is difficult to read on red).

These problems suggest that the color scheme for showing votes on the Summary/Review page should be changed. The design of the system we used mimics the interaction design of voting systems that were on the market at the time of the study. Considering the problems that we observed with the red blocks on the Summary/Review page along with the problems of registering touches and the awkwardness of having to deselect to change votes, it seems unlikely that better instructions on the page would solve all the issues for all voters. In Part 5, Recommendations, we recommend changes to this method of showing votes.

Did plain language make a difference?

Yes, but…

From the title onward, participants overwhelmingly found the instructions on Ballot A to be too terse and the instructions on Ballot B to be useful and helpful. The message about the red blocks was more informative and the step-by-step instructions on how to change votes helped participants understand what they needed to do, even when they were having difficulty registering their touches on the touch-screen.

A minor problem on Ballot B was the placement of two green buttons next to each other. We discuss that further when looking at the preference data for this page and recommend changes in the placement of the green button in Part 5. Recommendations.

The problem of helping people understand undervoting is much more important than the green button. Even with the greater explanation on Ballot B, some participants were confused when they tried to interpret the explanation "Pay attention to any red area. Red means you did not vote or did not vote for as many candidates as you could." As we said above, the red color was so alarming to many participants that they did not understand that it was acceptable to leave a contest with fewer than the maximum number of candidates selected.

While the user interface certainly needs to be toned down to be less alarming, alternative wording on the paragraph about red messages may also help. A different visual design and alternative wording for instructions would have to be tested together to determine whether the combination assists voters better than the combination used in this study.

Part 4
Discussion: Which ballot did participants prefer in a page-by-page comparison?

In this section, we look in detail at the preference data from the page-by-page comparison of 16 comparable pages in the two ballots.

Each ballot consisted of 26 screens (here "pages"), including the screen for writing in a candidate. The 16 pages that we included in the page-by-page comparison were

- Instructions to Voters/How to Vote
- Straight-party Vote/Straight-party Voting (page with parties listed)
- Straight-party Vote/Straight-party Voting (page with message after voting straight-party; choose whether to review and change or bypass / skip)
- President and Vice President
- Registrar of Deeds (Write-In Candidate/Write in a candidate's name)
- Write-In Instructions/Write In a Candidate (page for actually writing in a name)
- County Commissioners
- City Council
- Non-partisan offices (separator page before the non-partisan contests)
- Water Commissioners
- Retention Question/State Supreme Court Justice
- Amendment H/Amendment K (response options differ; "good" idea)
- Measure 101 (response options differ; "bad" idea)
- Ballot Summary/Review Your Choices
- Confirm
- Thank You

The 10 pages that we did not include in the page-by-page comparison were

- Opening page with precinct identifier and date
- Six pages of "vote for one" partisan contests
- One page of a "vote for one" non-partisan contest
- A second retention contest (State Supreme Court Associate Justice)
- A third amendment/measure page

We have organized this section moving from the page that scored the highest percentage preference for Ballot B to pages where, in fact, more participants preferred the version on Ballot A than on Ballot B.

Page	Thank you (last page in each ballot)			
Preference Thank you	Ballot A	2.2%	1	of 45
	Ballot B	97.8%	44	of 45

Wording on page – Ballot A

> # Thank you

Wording on page – Ballot B

> # Thank you
>
> Your vote has been recorded.
> Thank you for voting.

Participants' comments on the difference in instructions

The overwhelming reason for selecting B was that B told participants that their vote had been recorded:

A8:   It's courteous, telling you it's recorded.

B23:   It gives you some assurance.

B25:   It makes you feel good. You feel better leaving. You know what happened.

Some focused on the fact that it brought closure to the task:

A7:   You know it's gone in. My vote went somewhere.

B17:   You know it's in the system.

B29:   Gives a confirmation that my vote has been recorded and it's a done deal.

B31:   Tells you your vote has been recorded, letting you know you're locked in.

The one person who chose A felt that the Thank You by itself was "sufficient."

What this tells us about the value of plain language in ballots

In the tension within plain language between being specific and being short, our choice to be specific was supported by 98% of the participants. Notice that, in being specific in Ballot B, we followed the plain language guidelines of writing short sentences, using common words, and addressing the voter directly. In this case, we did use a passive sentence: Your vote has been recorded. We did that because in the active equivalent we would have had to name "the system" or "the machine" as the actor. In this case, the short, five word sentence worked well even in the passive.

Page	Instructions to Voters / How to Vote		
Preference			
Instructions to Voters	Ballot A	11.1%	5 of 45
How to Vote	Ballot B	88.9%	40 of 45

Wording on page – Ballot A

> # Instructions to Voters:
>
> Press the box of the Candidate for whom you desire to vote; yellow will appear in the box. The voter must re-touch the selected item to deselect it first in order to change a vote or in case of a mistake; then the voter touches the new Candidate of choice. Press Write-In to vote for a candidate who is not already listed on the ballot. On the Write-In screen, you must type the person's name and then press Accept (or press Cancel if you change your mind). Moving ahead is accomplished by touching the word Next; moving back by pressing Back.

Wording on page – Ballot B

How to Vote:

To vote for the candidate of your choice, touch that person's name. It will turn yellow.

To write in a candidate: To vote for a person who is not on the ballot, touch **Write in a candidate's name**. You will get more instructions on how to complete your write-in.

If you make a mistake or you want to change a vote, first touch the yellow box you no longer want. That box will turn gray. Then touch the choice you do want.

Participants' comments on the difference in instructions

Almost 90% of the participants preferred the plain language instructions that started the process.

Many had negative comments about the single paragraph on A.

A5: It's a lot of reading.

B17: When I first read it, I was overwhelmed. I had to read it three times. There was so much for me to remember.

C35: A big block of words. I don't think I read it all.

C39: I didn't read it through and had difficulty with straight-party because of that.

Many felt that B was less wordy and, therefore, easier to read through.

A8: It's less wordy and highlights words for reference. I'm a scanner.

B15: It's more concise, tells what you need to know.

B28: It's clearer, you get it faster.

C43: You want to get in and get out quickly.

They recognized and were positive about the three separate sections on B.

A11: It's simpler to see the directions and know immediately what to do.

B21: It's broken down more. This is what you do, step by step.

C32: I like the way it's broken up.

For many, the bold was used effectively on B.

A3: The important points are in bold. I like that.

C33: It helps the way it's set up; looks like less to read with shorter sentences. And I like the bold.

C39: I prefer B because of the bold.

What this tells us about the value of plain language in ballots

These two pages illustrate the difference between traditional and plain language perhaps better than any other set of pages in this study. The very high preference for the Ballot B page and participants' reasons for their preference show that they saw and understood the difference.

Page	Ballot Summary/Review Your Choices		

Preference

Ballot Summary	Ballot A	11.1%	5 of 45
Review Your Choices	Ballot B	88.9%	40 of 45

Wording on page – Ballot A

Ballot Summary

Press the candidate name or contest title to return to that part of the ballot. Contests that have red messages have not been completely voted. If satisfied with current selections, press "Next." Otherwise, press "Back"

Press here to cast your ballot now

Wording on page – Ballot B

Review Your Choices

This screen shows everything you voted for. Review it carefully.

Pay attention to any red area. Red means you did not vote or did not vote for as many candidates as you could.

To make changes:

1. Touch the race you want to change.

2. At that race, if you have selected something before, touch the choice you do not want. That box will turn gray.

3. Then touch the choice you want.

4. To return to this screen, touch Return to Review.

If you are ready to cast your ballot, touch Ready to Cast Ballot.

Ready to Cast Ballot

Participants' comments on the difference in instructions

Participants gave the plain language instructions at the end of the process the same very high preference as the plain language instructions at the beginning of the process – almost 90%.

The paragraph on A confused participants.

A11: I think I would have to click ? [help] for A.

A14: This is very, very confusing.

B21: This is not helping me very much.

B27: On A, this didn't tell me that I had to deselect first.

Some participants said that the page title on B and the first paragraph encouraged reviewing for them in a way that A did not.

A10: B has a better heading, more at a casual level that engages me more.

B26: Summary [A] says it's done; B is telling me to review, so I would.

For most participants, the more detailed information was helpful.

A4: More detail about making changes.

A8: Goes more in depth on how to do it properly.

B21: Coming to the end of me casting my ballot, I'd rather have a review telling me what to do.

C45: I needed all this information to figure out what to do. Even though it's a lot, I needed it.

And participants liked the instructions set out in steps.

A6: Steps to make changes. I could get to them quicker and do them more easily enumerated like that.

B30: I like the breakdown and the numbering.

C44: I'm a bullet-point type of person.

Several participants said that the green button appearing twice was confusing. (The one in the sentence is not live; it's just to show what it looks like. The second one is an active button.) In Part 5, Recommendations, we suggest changing the non-active button.

Note: The five participants who preferred A for this page were not the same as the five participants who preferred A for the instructions at the beginning.

What this tells us about the value of plain language in ballots

The very high preference scores for this set of pages once again show that participants recognized and valued plain language. In addition to the plain language guideline to be specific and give people what they need, the Ballot B version of this page exemplifies other plain language guidelines, including

- When you want people to act, focus on verbs rather than nouns. (Review Your Choices instead of Ballot Summary)

- Break up the information.

- When giving people instructions that are more than one step, give each step as an item in a numbered list.

(See Part 5, Recommendations, for a complete list of the plain language guidelines.)

Page	Confirmation		
Preference Confirmation	Ballot A	13.3%	6 of 45
	Ballot B	86.7%	39 of 45

Wording on page – Ballot A

> # Confirm
>
> The voter may now press "Confirm" to finish casting the ballot.
>
> The voter may press Return to ballot if the voter desires to make any more changes or selections.

The buttons on the bottom of the page on A were:
Return to Ballot   Confirm

Wording on page – Ballot B

Confirm

Are you sure you have finished voting?

Note: Once you touch Cast Vote , you will not be able to make any more changes.

If you want to make changes, touch Return to Ballot .

If you are ready to cast your ballot, touch Cast Vote.

The buttons on the bottom of the page on B were:
> Return to Ballot   Cast Vote

Participants' comments on the difference in instructions

Many aspects of the two versions drew comments from participants.

Although B has more words, participants felt the information is necessary.

> A2:   I like the specificity [of B].

> A13:   [B is] more specific; gives more information; tells you what to do
> if you're not sure; how to change.

> B22:   [B] is a much better explanation.

> C41:   It might be excessive, but it's really clear.

Some participants liked that it asked them if they were sure.

> B21:   It's like, "Wait a minute! Are you sure…"

> C32:   Asking me: "Are you sure?" I like that.

> C42:   I like that ["Are you sure?"]. It makes you stop and think one more
> time.

Others focused on the words that indicated how final the options here were.

> A12:   [B is] better – that you're about to reach the point of no return.

> B20:   [B] lets you know that once you cast your vote, you can't make more changes.

> C36:   If you select Cast Vote, you cannot change anything. A doesn't say.

A few noticed the order of the information and liked that the warning comes first.

> C45:   It warned you about not being able to make changes BEFORE it gave
> you instructions on how to make changes.

Participants also reacted positively to the layout.

C39:   There's more information. I like the way it's spaced out.

Several mentioned that it showed the buttons they could choose and made them look like buttons. (This reaction is interesting because it contrasts with the negative reaction to the two green buttons on the previous Review Your Choices page. It may be that on this page the non-live buttons in the instructions were not in color. It may be that the actual buttons were on the bottom of the screen not right next to the instructions as the green button was on the Review page.)

A12:   The bold and rectangles are giving you good options as to what to do.

C34:   Shows you what buttons to click rather than a sentence.

Participants also preferred the Cast Vote button to the Confirm button.

B17:   [B is] more assertive because of Cast Vote.

B29:   I see Cast Vote. That was it.

C32:   Cast Vote makes me feel like I'm actually pulling the lever.

The few participants who preferred Ballot A for this page liked that it was shorter. They did not feel that they needed the more explicit instructions.

What this tells us about the value of plain language in ballots

Again, participants recognized and valued plain language. The Ballot B version of this page differs from the Ballot A version through these plain language guidelines:

- Address the reader directly with "you" or the imperative ("Do x.").
- Put context before action, "if" before "then."
- Put information in the order that voters need it. Don't tempt voters to irrevocable actions before explaining the other options.

Page

**Write In Instructions / Write In a Candidate
(page for actually writing in a name)**

Preference

Write-In Instructions	Ballot A	13.3%	6 of 45
Write In a Candidate	Ballot B	86.7%	39 of 45

Wording on page – Ballot A

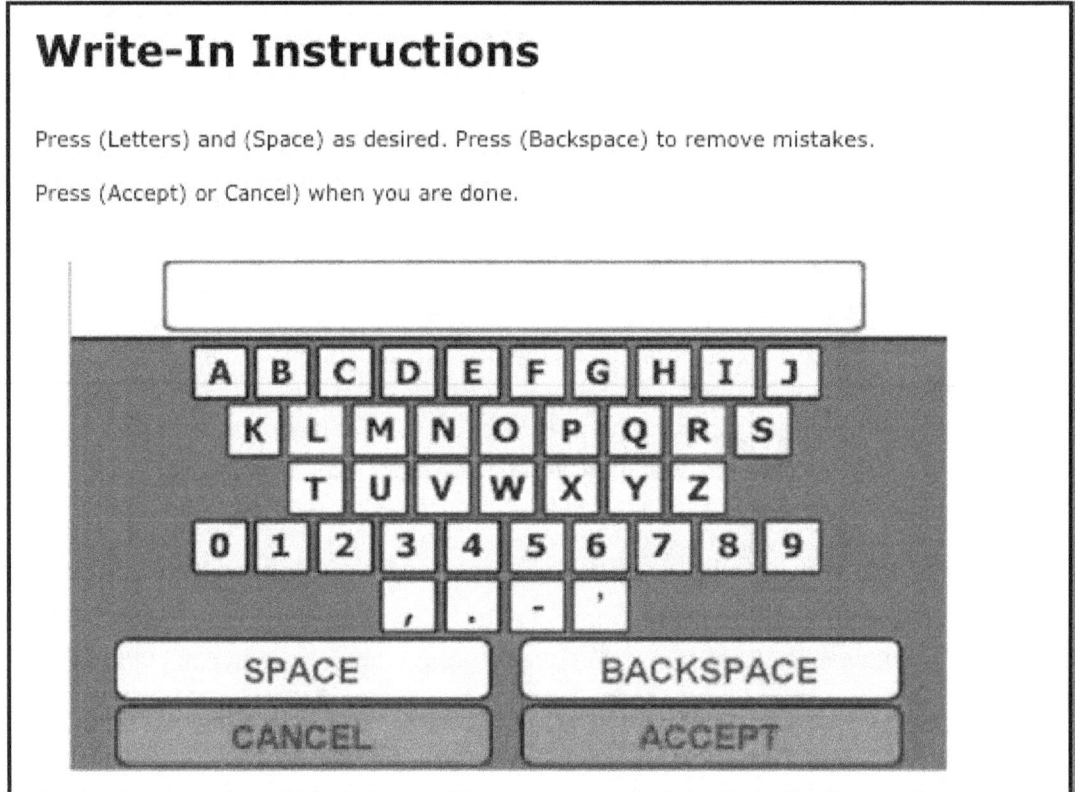

Wording on page – Ballot B

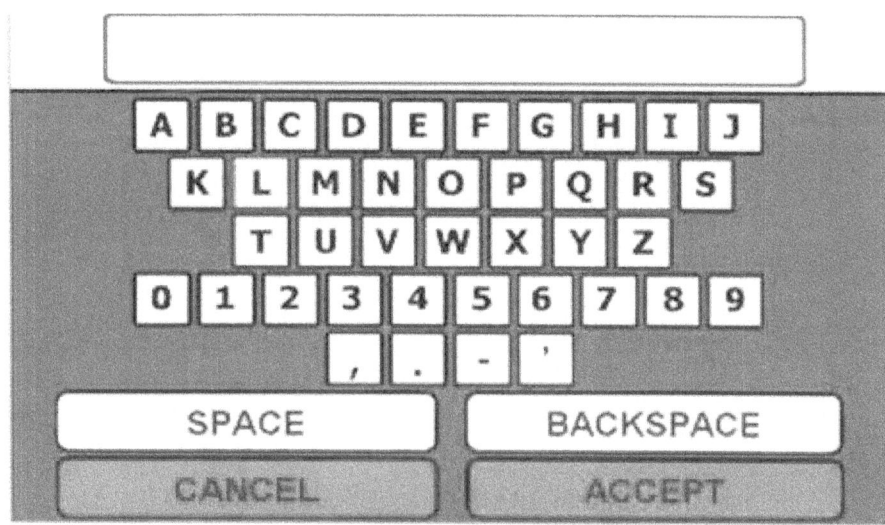

Write In a Candidate

Use this screen to vote for a candidate who is not on the ballot.

Do not write in someone whose name is already on the ballot for this race.

To write in a candidate:

- Type the person's first and last names.
- Put a blank space between the first name and the last name by touching SPACE.
- To erase, touch BACKSPACE.

To complete the write-in, touch ACCEPT.

If you change your mind, touch CANCEL.

| A | B | C | D | E | F | G | H | I | J |

| K | L | M | N | O | P | Q | R | S |

| T | U | V | W | X | Y | Z |

| 0 | 1 | 2 | 3 | 4 | 5 | 6 | 7 | 8 | 9 |

| , | . | - | ' |

| SPACE | BACKSPACE |

| CANCEL | ACCEPT |

Participants' comments on the difference in instructions

Almost all participants preferred the more explicit instructions on Ballot B with the buttons in color to the sparse instructions on Ballot A.

A5: [B is] more user-friendly; it tells you what to do if you make a mistake.

A10: [B] has more information on how to do it; [with A] you might feel a little flustered without the additional information.

B18: [A] doesn't really explain it to you. [On B, I like the] color buttons and detail.

B26: [B] It's more in detail; it tells you what it really wants you to do.

C33: I like that [the screen on B] uses color in the instructions to coordinate with the color on the buttons,

C38: [B is] more accurate in what they want you to do.

Page	Non-partisan offices (separator page before the non-partisan contests)

Preference	Ballot A	15.6%	7 of 45
Non-partisan offices	Ballot B	84.4%	38 of 45

Wording on page – Ballot A

Non-partisan offices

Wording on page – Ballot B

Non-partisan offices

If you voted a straight-party ticket, you have not voted for any race from here to the end of the ballot.

Participants' comments on the difference in instructions

Again, most participants opted for the page with an explanation. In this case, several participants remembered that the page with just a title had stumped them when they were voting. They thought that the screen had perhaps not fully refreshed and wondered why it was blank.

A9: Why is [A] here? It tells me nothing. I thought there was something wrong, that the machine was malfunctioning.

A14: I don't know why this one [A] is blank. Must have made a mistake because it's blank.

B20: There's nothing in the middle [on A]. [On A], I had to ask a question. I was clueless.

B23: I thought [A] might be an error.

B25: [A] is just blank. What am I supposed to do? [B] is more specific.

B31: [A] tells me nothing. I want to draw a picture [on the blank screen].

C32: When you stand in the booth and there's a blank page, you wonder.

C33:   I was confused when I got here [on A]. I was wondering if something was going to load up here.

C39:   [A has] absolutely no information about what you're looking at.

C42:   [B] tells you that you have more to come. [A] is blank so you would assume you're done.

C44:   [On A] I just stared. Do I do something now? I didn't know if the page had fully loaded.

Many participants who preferred B did not specifically comment on the explanatory sentence, but a few participants told us how the sentence on Ballot B explained the situation.

A5:   [B] gives you information about if you chose a straight-party ticket.

B17:   [B] explains it. It means you don't have to vote for those people because you already picked them.

The explanation on B, however, confused some participants. Although some of those participants chose B over A because it was not just a blank page, others preferred A. The sentence on B did not help participants who did not know the terms "partisan" / "non-partisan" or who did not understand straight-party voting. (See Part 3, Discussion: Where did people have problems?)

A12:   [Considering B] I've done what? Or I haven't done what?

A14:   I don't quite understand [B]. I assume your straight-party ticket was behind what you did. It's confusing. It should say why I haven't voted from here to the end.

B19:   I didn't like [B]; it was confusing.

B22:   I didn't understand A at all because I don't know what "non-partisan" means. B is better because it says something, but I would change what it says.

B29:   [B] made me wonder if I did the right thing.

C36:   [B is] kind of weird, confusing. I think you had to go through all those people anyway.

C43:   I wasn't 100% sure what [the sentence on B] meant.

What this tells us about the value of plain language in ballots

The difference here, which participants recognized, was the need to be specific and give people information they need. However, participants' comments also tell us that even Ballot B was not in plain enough language for some voters.

Page	Amendment H / Amendment K		
Preference Amendments H and K	Ballot A	15.6%	7 of 45
	Ballot B	84.4%	38 of 45

Page	Measure 101		
Preference Measure 101	Ballot A	17.8%	8 of 45
	Ballot B	82.2%	37 of 45

Wording on both pages – Ballot A

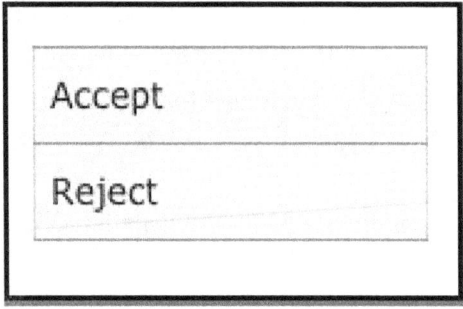

Accept

Reject

Wording on both pages – Ballot B

For

Against

Participants' comments on the difference in instructions

We showed participants two of the three pages of amendments and measures.

We had told them that they were to vote as if they thought this particular amendment (H on Ballot A; K on Ballot B) was a "good" idea. We had told them to vote as if they thought that this measure (101 on both ballots) was a "bad" idea.

We wanted to get their preference for the wording of the options and to see whether that preference would change depending on whether they were voting "accept"/ "for" compared to "reject" / "against."

Only one participant changed preferences, choosing "for" and "reject." All other participants stuck with their preference for one pair over the other.

In the performance data, neither set of words caused problems. Most of the errors came from participants forgetting that we had told them how to vote on these items. Many participants took the time to read the amendments and measures and, in a few cases, voted based on their own responses and not on our instructions. While we were listening to participants vote, we heard all able to read both sets of words easily.

In the preference data, which we are discussing in this part of the report, however, you see that a very large majority of participants (more than 80%) preferred "for"/"against" over "accept"/"reject." Their reasons were based both on the simplicity of the wording and on the sense that "for" and "against" better matched their relationship to amendments and measures.

A3: I'm used to the words [for/against].

A6: It's more appropriate to say [for/against].

A11: [For/against] cuts to the chase a bit quicker.

B15: [For/against] are simpler words.

B17: [For/against] sound better. They empower me more.

B21: [For/against] are easier to comprehend. You say you're for or against something.

B27: [For/against] is more of how you would feel about it.

B31: [For/against] are more like an election format.

C33: [For/against] In the general population, that's what people say.

C39: [For/against] are more direct.

C43: [For/against] are more common language.

One of the participants who opted for A here would have preferred "yes"/"no" rather than either choice that we offered.

What this tells us about the value of plain language in ballots

On these pages, the clear preference for the shorter, simpler, more common, everyday English words is an obvious indication of the value of using plain language on ballots.

Page	Retention Question / State Supreme Court Chief Justice

Preference

Retention Question	Ballot A	22.2%	10 of 45
State Supreme Court Chief Justice	Ballot B	77.8%	35 of 45

Wording on page – Ballot A

Retention Question

State Supreme Court Chief Justice. Shall Robert Demergue be retained?

Yes
No

Press Yes if you want the official to be retained in office.

Press No if you do not want the official to be retained in office.

Wording on page – Ballot B

State Supreme Court Chief Justice

Keep Esther York as State Supreme Court Chief Justice?

To keep Esther York, touch Yes.

To not keep Esther York, touch No.

Yes
No

Participants' comments on the difference in instructions

Although, of course, we always referred to the ballots as just "A" and "B" and never used the words "plain language" when talking with participants, it is clear that language was a major factor in people's preference here.

A4:   "Keep" sounds short and sweet versus "retain."

A12:   [A] is wordier. The explanation [on B] is far more simple.

B15:   [B] is more concise.

B25:   [B] is easier, quicker to read.

B29:   [B] just told the story.

C32:   "Retain" makes me think just a few second longer. "To keep" yes, yes I do [want to keep her]. Like I'm thinking 30 seconds less.

C36:   "Keep" is easier to understand than "retain."

C38:   I wasn't sure what "retain" or "retention" meant.

C42:   "Retention," "retain" someone might not understand. "Keep" is simpler, easier.

In this contest, the titles also differed. In the plain language ballot, we changed from the title, Retention Question, which often appears on ballots, to the name of the office, State Supreme Court Chief Justice. Participants commented on that difference.

A10:   [B] tells me what my topic is.

B16:   [The title of A] threw me off a little bit. It should have the race at the top.

B28:   I was a little confused because of the title on A.

C33:   "Retention" is about "remembering." I like the set up [of B] better. It let's me know what I'm doing right off the bat.

C41:   [B] let's you know what race you're voting for, not just that they're being retained.

A few participants focused on the fact that in the plain language version we also repeated the person's name.

A3:   Who is "the official"? That's too vague.

A9:   [B] is simpler. It restates the person's name as opposed to just "the official."

A few participants preferred the more formal language of A, but they were a small minority.

B24:   They're about the same, but [A] looks more official.

C37:   "Retain" is stronger, more professional.

Although no one mentioned it in the interview, in our observations, we saw that many participants did not read or look at the instructions on A that come under the choices rather than in the left column. This is typical of behavior that others have observed in people working with forms.

What this tells us about the value of plain language in ballots

Although the percentage of participants who chose Ballot B was slightly lower on these pages than on those we have discussed before, we still see more than ¾ of the participants focusing on the shorter, simpler, more common, everyday plain words and the more specific information of Ballot B.

Page	Registrar of Deeds		
Preference Registrar of Deeds	Ballot A	28.9%	13 of 45
	Ballot B	71.1%	32 of 45

Wording on page – Ballot A

> # Registrar of Deeds
>
> Vote for one.
>
> | |
> |---|
> | Laila Shamsi
Tan |
> | Write-in Candidate |

Wording on page – Ballot B

Registrar of Deeds

Vote for one.

Lillian Cohen
Lime
Write in a candidate's name

Participants' comments on the difference in instructions

Here, we were focusing on the choice for "Write-in Candidate" compared to "Write in a candidate's name." The Ballot A version is a noun phrase, matching the names of the other candidates. The Ballot B version is a verb phrase, matching the action the voter takes. The Registrar of Deeds page was where participants used this option. Our instructions to participants were:

> Ballot A: Even though you voted for everyone in the Tan party, for Registrar of Deeds, you want Herbert Liddicoat. Vote for him.

> Ballot B: Even though you voted for everyone in the Lime party, for Registrar of Deeds, you want Albert Sterner. Vote for him.

To many participants, the difference in wording was not compelling. They understood both and also understood what our instructions were asking them to do. (That is, when they saw that the candidate they were to vote for was not listed on the screen, they realized they needed to select the Write-in Candidate / Write in a candidate's name box.)

The difference in language here was greatly overshadowed by the problems people had either getting to this page (because they chose the wrong option after voting straight-party) or remembering that they had to deselect the party's candidate before being able to choose the write-in option. (See Part 3, Discussion: Where did participants have problems?)

Nevertheless, when pressed to see the difference and comment on it, more than 70% chose the verb phrase, focusing on the action and on the word "name."

A9: [B] tells you that you have to know the name.

A11: [On A], you don't get a good idea that you can write in a candidate.

A13:   [I prefer B] because of "name" -- that's a little more information.

B17:   [B] because it's giving the action.

B19:   [B] to be as specific as possible.

C35:   [The word "name" on B] really does make a difference. You wouldn't think it would, but it does.

C36:   [B] is easier to understand; it's more of an instruction.

C39:   [B] because it's more like a command. The word "name" is important.

C45:   The verb instruction is a reminder.

Some of those who preferred A were either worried that people would try to write in the candidate's name in the box on this screen – or had actually tried that themselves. (Again, we are seeing the reaction to the language confounded by the problem people had using the interface. When they were not able to select the write-in box [because they had not deselected the already-chosen candidate], several participants tried to use the stylus to handwrite the person's name into the box or elsewhere on the Registrar of Deeds page.)

B15:   I would write it in [the box on B] and I would be wrong.

Others preferred A because it is shorter.

B16:   [I prefer A] for the sake of simplicity.

C34:   [A] is shorter.

What this tells us about the value of plain language in ballots

More than 70% of the participants chose the verb phrase over the noun phrase. In this case, they may not have been aware of the plain language guideline they were reacting to but it is a useful one: When you want people to act, use verbs rather than nouns.

Pages   **Straight Party Vote/Straight Party Voting**

This is a set of two pages with the same title. On the first, voters either selected a specific party or decided not to vote straight-party. If they selected a specific party, they saw the second page. That page reminded them of the party they selected and gave them a choice of reviewing (and potentially changing) their party-based choices or bypassing/skipping all party-based contests.

Preference both pages   Ballot A   35.6%   16 of 45
                             Ballot B   64.4%   29 of 45

Wording on first Straight Party Vote page – Ballot A

Straight Party Vote

Vote for not more than one.

You may choose to vote a straight-party
ticket or vote each partisan contest.

Gold
Orange
Tan
Yellow

Wording on second Straight Party Vote page – Ballot A

Straight Party Vote

You have chosen to vote for ALL the candidates of this party:

Tan

Press here to review or change partisan selections

OR

Press here to bypass partisan contests

The buttons shown in the screen shot above are active. Voters are expected to use one or the other to take an action on this page.

Wording on first Straight Party Voting page – Ballot B

Straight Party Voting

You can vote all at once for all the candidates of one political party for all the races where candidates belong to a specific party. (This is called a straight-party ticket).

If you want most candidates from one party but some candidates from another party, you can vote straight party here and change your vote later at a specific race.

To vote straight party, touch the party name and then touch Next.

To not vote straight party, just touch Next.

Aqua
Lime
Purple
Silver

Wording on second Straight Party Voting page – Ballot B

Straight Party Voting

You have selected all of the candidates from the Lime party for all party-based races. (A straight-party ticket.)

If you want to change your vote in any party-based race, touch Next. When you get to the race you want to change, first touch the yellow box you no longer want. It will turn gray. Then touch the choice you do want.

If you want to keep just the Lime choices in all party-based races, touch Skip. You will go to the first non-party-based race.

The buttons at the bottom of this page are:
Back   ?   Next   Skip

Participants' comments on the difference in instructions

The preference statistics for these two pages are identical: 16 participants chose A; 29 chose B. However those numbers hide the fact that some participants were not consistent in their preference across the two Straight Party Vote/Straight Party Voting pages. When we look at the two pages as a set we see:

Prefer BB: 23 participants

Prefer AA: 9

Prefer AB: 7

Prefer BA: 6

The 23 participants who chose B for both pages focused on the more in-depth explanation and the clearer wording (not using "partisan").

A3:   Page 1: [A] is short but confusing; [B] has more detail and I got confused on A [without that detail]. Page 2: [A] had me totally confused. I had no idea what "partisan" meant.

A8:   Page 1: [B] has more detail of what you're doing. Page 2: [B] goes step by step, tells you how to do it properly so you're not confused about what to do.

B17:   Page 1: [On A] I didn't understand that there were options. I just thought it was a description. [B] tells you what to do. Page 2: [B] makes sense for me.

B20:   Page 1: [B] tells you what is a straight-party ticket because I don't know what that is. Page 2: [On A]: I don't know what none of these meant.

C35:   Page 1: [A] has less to read but not much information. [B] is way easier. Page 2: [On A] I didn't understand what they were talking about [pointing to "partisan"].

C39:   Page 1: [A] did not give much information about why you would vote straight-party or how to change. On [B] I knew from the first page that I could change later. Page 2: I was confused [on A] about what the selection meant, so I just clicked whatever seemed to make sense and realized later that's not what I wanted.

The nine participants who chose A focused on the fewer words.

A6:   Page 1: [A] is straight to the point; [B] is too wordy. Page 2: [A] has less for me to read.

A11:   Page 1: [A] is simpler, easier; [B] is a lot to throw at you. Page 2: [A] is more straightforward.

B15:   Page 1: [A] is less words. Page 2: [A] has less [to read].

Seven participants switched from A to B. They liked the few words on A, but then got confused on B and liked having more words because the explanation helped them.

A13: Page 1: [A] seems easier; [B] is too much information. Page 2: [B] gives you details, which is a good thing.

B28: Page 1: [A] is simpler, less writing. Page 2: [B] explains it.

C32: Page 1: [A] has short sentences and gets right to the point. Chances are, I wouldn't have taken the time to read more than about two or three sentences [of B]. Page 2: [A] threw me for a loop. It's a wasted page. I wouldn't consider this because I didn't know what lay ahead. [B] was quick and fast.

C42: Page 1: [A is] simpler, shorter, easier to understand.
Page 2: People are not going to understand [A].
[B is] simpler, more information.

Six participants switched from B to A. They liked the in-depth explanation on the first Straight Party Voting page on B but then didn't need or didn't like the in-depth explanation on the second B page. These may have been people who understood the concept of "straight-party" and "partisan" contests.

A1: Page 1: [A] doesn't tell you what you can do; [B] tells you exactly what you can do. Page 2: [A] is plain and simple; [B] is too much.

C37: Page 1: [On A, I though I] can't do straight-party because I won't be able to change votes later, so I prefer B because it lets you know you may change some of your votes. Page 2: The buttons for Next and Skip [on B] are confusing. I almost did Skip instead of Next. [A is] more straightforward.

C40: Page 1: [B] tells you more about going back to change parties later.
Page 2: [A] is straightforward. [B] gives details it already said earlier.

Some people suggested changes to the wording or wanted a compromise: the explanations of B but with fewer words, the shortness and direct buttons of A but without the words "partisan" and "non-partisan."

What this tells us about the value of plain language in ballots

Again we see fewer words competing with a clear, well-written explanation, broken into sections. The preference data, and especially the reasons participants gave for their preferences, reinforce the point that plain language is not just using few words. Plain language is saying what must be said to help voters be successful while giving those necessary messages in short sentences, separated into short paragraphs, with words voters know, speaking directly to them, and so on.

Although almost 2/3 of our participants preferred the Ballot B page for each of these two pages, we realized from their behavior and their preferences that we could have done a better job at implementing the plain language guidelines on the Ballot B version of the second Straight Party Voting page. In Part 5, Recommendations, we suggest changes to this page of Ballot B.

Pages
County Commissioners
City Council
Water Commissioners

The issue on each of these pages was the wording of how many people a voter could select. The Ballot A pages used the formula: "Vote for no more than x." The Ballot B pages spelled out the numbers from 1 to x.

Preference

County Commissioners	Ballot A	51.1%	23 of 45
	Ballot B	48.9%	22 of 45
City Council	Ballot A	53.3%	24 of 45
	Ballot B	46.7%	21 of 45
Water Commissioners	Ballot A	46.7%	21 of 45
	Ballot B	53.3%	24 of 45

Wording on County Commissioners page – Ballot A

Vote for no more than five.

Wording on County Commissioners page – Ballot B

Vote for one, two, three, four, or five.

Wording on City Council page – Ballot A

Vote for no more than four.

Wording on City Council page – Ballot B

Vote for one, two, three, or four.

Wording on Water Commissioners page – Ballot A

> Vote for no more than two.

Wording on Water Commissioners page – Ballot B

> Vote for one or two.

Participants' comments on the difference in instructions

Participants who favored A usually said that it was shorter.

- A1: [B] is too much reading.
- A6: [B] is redundant.
- A8: [A] is quicker.
- B17: [A] is simpler.
- B20: [A] is shorter.
- C37: [On B] someone might stop reading after getting to "one."
- C44: [A is] all that's necessary.

Some thought listing out the numbers meant that you were to rank the candidates in an order rather than voting for them equally.

- A3: You don't need all the numbers on [B]. It makes you think you've got to rank it almost.
- A13: [On B] I wasn't sure if I had to vote giving them an order 1, 2, 3, 4, 5.
- B26: [said same thing]

Those who favored B liked the options listed out.

- A10: [B] has more information. I like the assurance that 1 is fine. [Listing the numbers out] better matches the way my brain thinks.
- B25: [B is] more specific.
- B30: I liked [B] because I wasn't sure. With [B] there's no gray area.
- B31: [On B] if I just want one, I can do that. It's letting me know I have a decision.
- C43: [B] is more explanatory. I understood it better.

Nine participants who favored B interpreted the version on A to mean voters must vote for that number.

A2:   [On County Commissioners A] you would have to vote for 5.

A5:   [A] makes you think you have to vote for at least 5.

A7:   [With A] I had it in my head that I had to vote for 5.

A10:   [On Water Commissioners A] I automatically voted for 2 because of what it said.

B25:   [With Water Commissioners A] someone could get confused that they would have to vote for 2.

B31:   [On A] I have to vote 5 even if I don't like them.

C34:   [On A] I wasn't sure if any less than 5 was okay.

C42:   [On A] Assumption: You have to vote for 5.

C45:   [A] says you *must* vote for 5.

One mentioned the lack of zero in the listed-out version on Ballot B and another interpreted the list without a zero option. And yet, these same people had no problem leaving a contest blank as we told them to in their first pass by the Water Commissioners contest.

A5:   [On City Council B] This means at least one but no more than four.

A7:   I don't like that [B] doesn't say "zero."

As you can see from these quotes, some people (like A7) had problems with both the A version (makes you think you must vote for that number) and the B version (doesn't say you can vote for zero).

Participants were not always consistent in their preference across these three pages.

Some chose the Ballot A version for the first two (no more than five; no more than four) and the Ballot B version for the Water Commissioners page (vote for one or two). Was this because listing out more than two numbers seemed excessive to them? Was it that "vote for one or two" is actually one word shorter than "vote for no more than two"?

B29:   [On County Commissioners] [A] means don't go past 5; you don't have to go up to 5; [B] is confusing. [On Waster Commissioners] [B] is better, no guesswork.

C32:   [On County Commissioners] [B] is taking up too much time. [On Water Commissioners] [B] is reminding me; it's more comfortable.

C39:   [After choosing A on the first two as "less wordy" although "I don't see that big a difference," on Water Commissioners chose B saying about A] I had no idea that I could pick one or two.

C40:   [After saying A was "perfect" on the first two and "you don't need all the numbers," chose B on Water Commissioners] [B] has fewer words and you still know what to do.

At least one went the other way. Was that because the Water Commissioners was the third multi-candidate contest and this participant had gotten the message by then?

B30:   [After having a strong preference for B on the earlier two contests] For some reason, that one [the A version of Water Commissioners] worked.

What this tells us about the value of plain language in ballots

On these three pages, both the A version and the B version could be considered plain language. The A version represents the typical formula seen on ballots, and we had heard anecdotally that some people interpret "vote for no more than x" to mean "I must vote for x." Therefore, we tried a more specific formula on Ballot B.

On this issue, however, our study did not have a clear finding. As we suggest in Part 6, Conclusions and suggestions for future research, how best to explain multi-candidate contests needs further study.

Page	President/Vice President		
Preference President/Vice President	Ballot A	55.6%	25 of 45
	Ballot B	44.4%	20 of 45

Wording on page – Ballot A

Vote for one.

(A vote for the candidates will actually be a vote for their electors.)

Wording on page – Ballot B

Participants' comments on the difference in instructions

This page received the highest preference score for the Ballot A version. Ballot A had *more* words and *more* explanation than Ballot B. The higher preference for the Ballot A version, therefore, is in keeping with the overall trend of our participants to favor explanations on the ballot pages. However, almost half of the participants either did not understand it themselves or were concerned that others might be more confused with the sentence than without it.

Participants who preferred Ballot A thought people would want the extra fact:

A9:   I'm not sure it's necessary, but in the interest of full disclosure, it's more accurate.

A11:   It's good to know.

B23:   It's important to maintain integrity.

B31:   [A] gives you more detail of who you're voting for.

C39:   It's better to have more information.

Participants who favored Ballot B thought people didn't care or need to know that extra fact. They rejected A with these words:

A12:   The less of that kind of information, the better.

B27:   I don't think it's really necessary.

B28:   You don't really need all that.

C35:   It's information I don't care about. It just confused me more.

C41:   If people don't know exactly how the system works, they may get confused.

C44:   I don't think I would have read it. It's frivolous.

Several participants were themselves confused by the extra statement.

A14:   What does that mean?

C42:   I don't know what that even means.

C43:   I'm not 100% sure what that means.

C45:   I'm not sure I understand this. Aren't I an elector? It implies that you're voting for a group of people that you don't know who they are.

What this tells us about the value of plain language in ballots

The participants' reactions to the extra instruction on this page of Ballot A reveal less about plain language than about the need for civics and voter education. While it is the candidates' names that appear on the ballot, there's another process going on that generates electoral votes in addition to popular votes. Participants' preference for the additional information suggests that many think voters should be made aware that this contest isn't as straightforward as it appears. However, the closeness of the preference data and participants' comments about the wording mean that both the issue of whether to include this statement and the clearest way to state it need further study.

Part 5: Recommendations for creating a ballot that voters can understand and use successfully

In this part, we give recommendations based on what we learned in this study.

We begin with the overall recommendation to use the plain language instructions from Ballot B with a few changes. We are specific about the changes we would make to the plain language instructions.

We then summarize the guidelines from which we created the plain language instructions for Ballot B.

As we explained in Part 3 where we gave details of pages with high error rates, we have recommendations that go beyond language. We give those next.

And we end with a very strong recommendation for usability testing of each specific ballot.

While our recommendations in this part are for a plain language ballot in English, we reiterate the point we made on page 1 of this report that plain language is relevant to all voters and all languages. Many jurisdictions prepare ballots in multiple languages. In Part 6, Conclusions and suggestions for future research, we recommend studying this issue for other languages as well.

Recommendation 1: Use plain language in instructions to voters

The success of Ballot B in both performance and preference strongly supports a recommendation that all ballots should follow plain language guidelines.

The guidelines here summarize what we did when we created Ballot B. You should find these helpful whether you are creating ballots for an electronic voting system or for paper.

Many of these guidelines appear with more details and examples in an earlier document that is available from NIST at http://www.vote.nist.gov/032906PlainLanguageRpt.pdf.

We have also created a handout of these guidelines as an appendix to this report.

What to say and where to say it

Be specific. Give people the information they need.

At the beginning of the ballot, explain how to vote, how to change a vote, and that voters may write in a candidate.

Put instructions where voters need them. For example, save the instructions on how to use the write-in page for the write-in page.

Include information that will prevent voters from making errors, such as a caution to not write in someone who is already on the ballot.

On a DRE, never have a page with only a page title (such as the Ballot A page that just said Non-partisan offices).

Make the page title the title of the office (State Supreme Court Chief Justice rather than Retention Question).

Have voters confirm that they are ready to cast their vote with a Cast Vote button, not a Confirm button.

At the end, tell people that their vote has been recorded.

How to say it

Write short sentences.

Use short, simple, everyday words. For example, do not use "retention" and "retain." Use "keep" instead. For another example, use "for" and "against" for amendments and measures rather than "accept" and "reject."

Do not use voting jargon ("partisan" "non-partisan") unless the law requires you to do so. If the law requires these words, work to change the law. Instead refer to contests as "party-based" and "non-party-based."

Address the reader directly with "you" or the imperative ("Do x.").

Write in the active voice, where the person doing the action comes before the verb.

Write in the positive. Tell people what to do rather than what not to do.

Put context before action, "if" before "then." For example, To vote for the candidate of your choice, touch that person's name.

When you want people to act, focus on verbs rather than nouns. For example, Write in a candidate's name.

When giving people instructions that are more than one step, give each step as an item in a numbered list.

Do not number other instructions. When the instructions are not sequential steps, use separate paragraphs with bold beginnings instead of numbering.

Put information in the order that voters need it. Don't tempt voters to irrevocable actions before explaining the other options. (See, for example, the order of the information on the Ballot B Confirm page: a question, a note about consequences, an instruction on how to make changes, and then the irrevocable action described last.)

What to make it look like

Break information into short sections that each cover only one point.

Keep paragraphs short. A one-sentence paragraph is fine.

Separate paragraphs by a space so each paragraph stands out on the page.

Do not use italics.

Use bold for page titles.

Use bold to highlight keywords or sections of the instructions, but don't overdo it.

Keep all the instructions in the left column. Do not put instructions under the choices for a contest.

Do not use all capital letters for emphasis. Use bold. Write all instructions in appropriate upper case and lower case as you would in regular sentences. If the law requires you to use all capital letters, work to change the law.

Use a sans serif font in a readable type size.

Recommendation 2: Use our Ballot B language with these changes

Most of the specific language in Ballot B worked very well and produced the strong performance and preference differences that we described earlier in the report. We would, however, change a few specific elements to go even further in plain language as follows:

Changes to the How to Vote page

Reverse the second and third paragraphs on the How to Vote page. Tell people about the need to deselect before telling them about being able to write in a candidate's name.

Changes to the first Straight Party Voting page

Rather than having one Next button, have separate buttons for the two choices:

[Vote straight-party]   [Do not vote straight-party]

Changes to the second Straight Party Voting page

Rewrite the page to say:

> You have selected all of the candidates from the **Lime** party for all party-based races. (A straight-party ticket.)
>
> **To review or change your vote in any party-based race**, touch
> [Review party-based choices].
> You will go through the screens for the party-based race.
>
> **To keep all votes for the Lime candidates in party-based contests without reviewing them**, touch   [Keep all party-based choices].
> You will go to the first non-party-based race.

Make these the actual buttons – not at the bottom of the screen. The buttons placed this way on Ballot A worked for people.

Note the use of bold. The bold here acts like a heading would. It makes the two choices obvious. Many of our participants singled out the use of bold in a similar way on the How to Vote page as being an important reason for them to prefer that page.

Note that we have taken away the instructions on how to change a vote. We recommend saving those instructions for each contest page.

Note that we have put what will happen after each choice on a new line, so the line of instruction ends with the button.

Changes to all pages where voters select candidates or other options

Put the instruction about how to deselect a choice before selecting a new choice on every page where deselection may be needed.

Changes to the Review Your Choices page

Use a color other than red to indicate undervotes.

Consider a different color scheme; for example:

- Red to show races that were not voted at all

- Orange to show that the voter could have voted for more choices

- Green to show that the race was completely voted

A change in the color scheme would, of course, require a change in the instructions on the Review Your Choices page. In revising any instructions, we strongly recommend keeping the simple language, short paragraphs, color-coding, and step-by-step directions that characterize Ballot B.

For multi-candidate contests, indicate the maximum number the voter could have selected and the number voted for. For example:

> **County Commissioners**
> (You have voted for 3. You may vote for up to 5.)
> Name 1 (Party); Name 2 (Party); Name 3 (Party)

Name the party of each party-based candidate the voter voted for in all party-based races.

Remove the color from the Ready to Cast Ballot button in the instruction. Move the live green Ready to Cast Ballot button to the bottom and replace the Next button with the Ready to Cast Ballot button.

Recommendation 3: Put each contest and measure on its own page on a DRE

Another critical guideline that we followed with both ballots was to present each contest and measure on its own page. Other research has shown that this is vital. Many people miss the second race when one DRE screen has more than one race.[10]

---

10 Norden, Lawrence, David Kimball, Whitney Quesenbery, and Margaret Chen, *Better Ballots*, Brennan Center for Justice at NYU School of Law, 2008. Available at http://www.brennancenter.org/content/resource/better_ballots/

Recommendation 4: Consider removing straight-party options from ballots

This study showed that much as plain language can help, it cannot solve all problems in voting. Most of the errors that our participants made were related to straight-party voting and wanting to change some party-based races after voting straight-party.

As we described in detail in Part 3, Discussion: Where did participants have problems? many of our participants did not understand the concept of straight-party voting. They didn't know what that meant. They did not know the words "partisan" and "non-partisan." But even using "party-based" rather than "partisan" did not resolve the difficulty for all participants.

The concept of voting straight-party was difficult. And the concept of both voting straight-party and then changing some of the party-based races was even more difficult.

We recommend that states think about this issue and review their policies with consideration of our findings.

Recommendation 5: Do more with voter education materials

In addition to their problems understanding straight-party, many of our participants did not have a clear concept of the different levels of government. They mistook the U S Senate race for the State Senator race. They mistook the County Commissioner race for the City Council race. And this problem is made worse by the fact that, on a DRE, voters only see one contest at a time and do not know what other contests are coming later on the ballot.

The language on the ballot itself cannot compensate for this lack of understanding. Voter education before the election is needed. To be successful, voter education materials must also be in plain language, and the guidance in these recommendations for language and layout is relevant to all materials for voters.

Furthermore, sample ballots must look like the ballots that voters will use in the polling place.

Recommendation 6: Test ballots with voters before each election

Based on this study, we can strongly recommend the design and language of Ballot B (with the changes listed earlier in this recommendations section) for all ballots. However, no specific ballot for any specific election in any specific jurisdiction is going to have exactly the contests and measures that we included in Ballot B. Local election officials constructing ballots are going to continue to make choices and decisions on every page of every ballot whether the ballot is delivered on paper or electronically.

The best way to guard against disaster in an election due to ballot design or language is to have a few voters try out the ballot before the design and language become final. The methodology for having voters try out a draft is usability testing. With even a few (8 to 10) representative voters going through a ballot against a slate (the methodology we used in this study), serious issues such as voting against one's intent (the "butterfly ballot") or just not seeing a contest at all (the Sarasota County problem) will become apparent. The voters who try out the ballots must not be the same people who designed and defined the ballot. They must be voters from the community, not workers in the Election Department. People in the Election Department (even those who were not directly involved in designing and defining the ballot) may know much more about voting than typical voters and may know voting "jargon" that typical voters do not know.

Part 6: Conclusions and suggestions for future research

This study has shown that the language on ballots can help or hinder voters. Clear, plain, well-presented language reduces errors. Voters recognize the difference between traditional non-plain language and plain language. They overwhelmingly prefer simple, clear, plain language instructions on ballots.

Of course, this study had limitations. Our 45 participants represented a wide spectrum of voters – but not all voters. We had one outlier who voted an empty ballot and we do not know what portion of the population she might represent. Our focus was on direct recording electronic (DRE) voting systems; while those are widely used today, many people vote on paper not on DREs. We contrasted two ballots that had several contests and some measures – but not all combinations that occur in elections. We found that many of our participants did not have a good mental model of ballots or levels of government, but we tested only ballots, not other voter education materials. We tested only English-language ballots with English speakers. Even though some of our participants may not have been native speakers, they all read and understood English.

These limitations and our findings lead to the following suggestions for future research:

Future research 1: Test with low literacy voters

Our study plan focused on participants with high school education or less. We successfully recruited to that plan, and doing so allowed us to consider how education correlated with performance on ballots. We found that education did indeed correlate negatively – our participants with lower education levels made more errors (particularly on the traditional language ballot).

But we did not specifically test for low literacy; and our participants, even those with lower education levels, read competently. They had only a few problems pronouncing words as they read the directions we gave them and the words on the ballots.

If the low-education-level but fairly competent readers in our study had problems, readers with lower reading skills might have even more problems. Further research with low literacy voters would be worthwhile.

Future research 2: Investigate the prevalence of people who vote empty ballots

Further research on the problem of empty ballots might be extremely important. How many empty ballots are cast on DREs in typical elections? Are there more in precincts in certain geographic areas, ethnic communities, socio-economic areas? Are there many voters out there who are so unused to computers and to voting that they would do as our "outlier" person did? (People may vote an empty ballot on purpose, knowing exactly what they are doing. However, our participant did not vote an empty ballot on purpose. She said she had voted for specific candidates and measures.) What can be done to help people like this participant become the voter she wants to be? What types of training would help?

Future research 3: Test with older adults

In our study with participants ranging in age from 18 to 61, we did not find differences in voting behavior among the different age groups. We did not have anyone older than 61 among our participants. We would be very comfortable making the hypothesis that plain language would matter as much if not more to older adults, but further research with that population would be needed to test that hypothesis.

Future research 4: Test with other modalities (for example, audio) and with special populations

Our study did not focus on modalities other than text nor on people with specific needs. As we just said about older adults, we would be very comfortable making the hypothesis that plain language would work better than traditional language for voters who listen to instructions (audio) and for voters with cognitive issues. However, further research would be needed with those populations to test our hypothesis.

Future research 5: Test with other languages

Plain language is not just an English-language issue. Jurisdictions that prepare ballots in languages other than English must consider the value of applying similar guidelines for those languages. We believe that clear, simple, direct, specific wording and presentation of instructions helps all voters in all languages.

It is a known fact in other domains (translating manuals, for example) that plain language facilitates translation. For jurisdictions that start in English and translate into other languages, a plain language ballot in English should make translation faster, easier, and more accurate.

We suggest replicating this study with other languages. We suggest that jurisdictions that translate from English keep track of costs of translation for traditional, non-plain-language ballots compared to translation for plain-language ballots.

Future research 6: Apply what we learned to paper ballots

We studied the language for the pages of a touch-screen ballot. Much of what we learned applies to paper ballots as well. Paper ballots, however, operate differently both in terms of what a voter can do and in constraints on where and how instructions can appear. We also know that reading differs between screen and paper. A study comparable to the one we completed should be done with paper ballots.

Future research 7: Do a similar study on a ballot without a straight-party option

Many of the problems we saw came from our participants not understanding the concept of straight-party and especially of being able to both vote straight-party and change a party-based contest. How many of the errors that we saw even with our plain language Ballot B would go away if straight-party were not an option? Our hypothesis would be that voters would make fewer errors even though they would have more contests to go through; but that is an empirical question.

Future research 8: Find the best way to design, write, and deliver voter education materials

Many people besides our outlier showed that they had little concept of voting and the many types of contests in elections. What type of voter education is most effective in helping these people understand the process of voting, where contests come on ballots, what the different levels of government are, etc.?

Not surprisingly, lower education achievement correlated with higher error rates. What can be done in school before these people leave school to prepare them for participating in civic life, including voting?

Future research 9: Look into specific changes for specific issues that came up in this study

Although the plain language Ballot B was much more effective in helping voters than the traditional language Ballot A, people still had problems with some aspects both of the ballot and of voting.

Studies that look at some of these specific issues would be extremely useful. Here are three examples:

- **Deselecting** was not a natural behavior for many of our participants. They did not remember that to change a vote in a contest where the maximum number of people or options was already selected they had to first click on a candidate or option that they no longer wanted. Their natural instinct was to simply chose the new candidate or new option. We recommend further study of this issue.

- **Red as the color to indicate undervoting** made several of our participants so uneasy that they insisted on voting for the maximum number to remove them. We have suggested a different scheme for the Summary/Review page. Would our suggested scheme of red/orange/green solve the problem? Would some other way of indicating that intentional undervoting is okay work better than our suggestion? For example, would it help voters if we put a comment on every ballot page saying that "You may vote for fewer than x candidates"?

- This study resolved only some of the issues **concerning the best way to tell people how many candidates they could vote for** in a contest.

  - This study may have resolved the question: Does the instruction have to mention a possibility of voting for none? When we specifically told people to leave a Vote for no more than two / Vote for one or two contest unvoted, they did that. Therefore, telling people how many they can vote for does not seem to make them feel forced to vote in a contest – even though a few participants mentioned that as a possible problem with listing out all the numbers in a multi-candidate contest. (The red blocks on the Summary/Review page did make several participants feel forced to vote; but the instruction did not.)

  - This study found a successful way to give an instruction in a single-candidate contest. We did not vary this instruction between the ballots; both had Vote for one. None of our participants had a problem with the instruction. However, we do not know if giving that instruction with the numeral would have been better: Vote for 1. A study to test that difference would be useful.

  - This study did not resolve the issue of how to best express the instruction in a multi-candidate contest. Slightly more participants preferred the formula, Vote for no more than x to the formula, Vote for [each number listed out] in contests with four and five as the maximum, However, slightly more participants preferred the numbers spelled out when the maximum was two. Of course, in contests where the maximum is more than five, listing out the numbers would get unwieldy. Further research on this would be helpful, although other issues may be of greater importance because both formulas worked for most participants.

Future research 10: Add other aspects of the voting process that we did not include

Our ballots did not give participants a paper trail. Just having paper is not in itself enough to ensure that people will notice the paper, be able to read it and review it before casting their vote, understand what to do if they do not agree with the paper, and so on. The design, language, and usability of audit screens and audit paper are also important issues and should be researched.

Appendix 1

Participants

Demographic and Voting Experience Questions (with response numbers added as a right column)

P # _____

Personal Information	Responses
1. Gender	
☐ Female	**23**
☐ Male	22
2. Age	
☐ 18-21	7
☐ 22-25	5
☐ 26-30	8
☐ 31-40	6
☐ 41-50	10
☐ 51-60	8
☐ Over 60	1

Education	Responses
3. What is the highest level of education that you have completed?	
☐ Less than a high school degree	9
☐ High school	15
☐ Some college	12
☐ College degree	8
☐ Some advanced courses beyond college	1
☐ Advanced degree	0
☐ Other, please describe _____	0

OMB Control No. 0693-0043   Expiration 07/31/09

Voting experience   **Responses**

4. How many years of voting experience do you have?

☐   None   **6**

☐   Less than 2 years   **5**

☐   2-5 years   **5**

☐   6-10 years   **19**

☐   11-20 years   **9**

☐   More than 20 years   **1**

5. In the past two years, how many government elections (national, state, or local) did you vote in?

☐   0   **10**

☐   1-2   **22**

☐   3-5   **11**

☐   more than 5   **2**

6. Different areas in the United States have used different types of voting systems over the years.

Which, if any, of the following have you ever used to vote in a government election (national, state, or local)? Please mark all the types you have used.

☐   I have never voted in a government election.   **6**

☐   Mechanical lever – where the voter sets switches and   **15**
    pulls a big lever

☐   Optical scan – where the voter fills in a circle or oval or   **15**
    connects arrows to indicate a vote and then the paper
    is checked by a machine

☐   Paper and pencil – where the voter marks a paper and   **14**
    the paper is checked by a person

☐   Punch card – where the voter uses a device that punches holes   **11**
    in a voting card

☐   Touch screen – an electronic voting system where the voter   **23**
    touches a screen to vote

Other, please describe _____   **0**

OMB Control No. 0693-0043   Expiration 07/31/09

7. Before today, have you ever voted for a write-in candidate; that is, written the person's name on the ballot?

   ☐ Yes   **10**

   ☐ No   **35**

8. Which of the following items do you use regularly?

Please mark all the ones that you use.

   ☐ ATM machine   **38**

   ☐ Cell phone   **43**

   ☐ Computer   **40**

   ☐ Device to record from your TV (DVD recorder, VHS recorder, other)   **29**

   ☐ Digital Camera   **33**

   ☐ Self checkout at grocery or other stores   **38**

Appendix 2

Ballot A

Official Ballot

General Election Ballot
City of Montcrest
Union County, Nebraska

November 6, 2007

Congressional District X
Senate District Y
Assembly District NN
Council Districts 1, 3, 5

Next

Instructions to Voters:

Press the box of the candidate for whom you desire to vote; yellow will appear in the box. The voter must retouch the selected item to deselect it first in order to change a vote or in case of a mistake; then the voter touches the new candidate of choice. Press Write-In to vote for a candidate who is not already listed on the ballot. On the Write-In screen, you must type the person's name and then press Accept (or press Cancel if you change your mind). Moving ahead is accomplished by touching the word Next; moving back by pressing Back.

Back	Next

Straight Party Vote

Vote for not more than one.

You may choose to vote a straight-party ticket or vote each partisan contest.

Gold

Orange

Tan

Yellow

Back (?) Next

Straight Party Vote

You have chosen to vote for ALL the candidates of this party:

Tan

| Press here to review or change partisan selections |

OR

| Press here to bypass partisan contests |

President and Vice President

Vote for one.

(A vote for the candidates will actually be a vote for their electors)

Joseph Barchi and Joseph Hallaren	Orange
Alvin Boone and James Lian	Yellow
Adam Cramer and Greg Vuocolo	Tan
Daniel Court and Amy Bluhardt	Gold

Back   (?)   Next

US Senate

Vote for one.

Dennis Weiford Orange
Silvia Wentworth-Farthington Yellow
Lloyd Garriss Tan
Write-in Candidate

| Back | ? | Next |

US Representative

Vote for one.

Brad Plunkard
Orange

Bruce Reeder
Tan

Write-in Candidate

Back (?) Next

Governor

Vote for one.

Charlene Franz Orange
Linda Bargmann Yellow
Gerard Harris Tan
Barbara Adcock Gold
Douglas Alpern Independent
Write-in Candidate

Back (?) Next

Lieutenant Governor

Vote for one.

Chris Norberg
Orange

Luis Garcia
Yellow

Anthony Parks
Tan

Write-in Candidate

| Back | ? | Next |

Registrar of Deeds

Vote for one.

Laila Shamsi
Tan

Write-in Candidate

Back ? Next

Write-In Instructions

Press (Letters) and (Space) as desired. Press (Backspace) to remove mistakes.

Press (Accept) or (Cancel) when you are done.

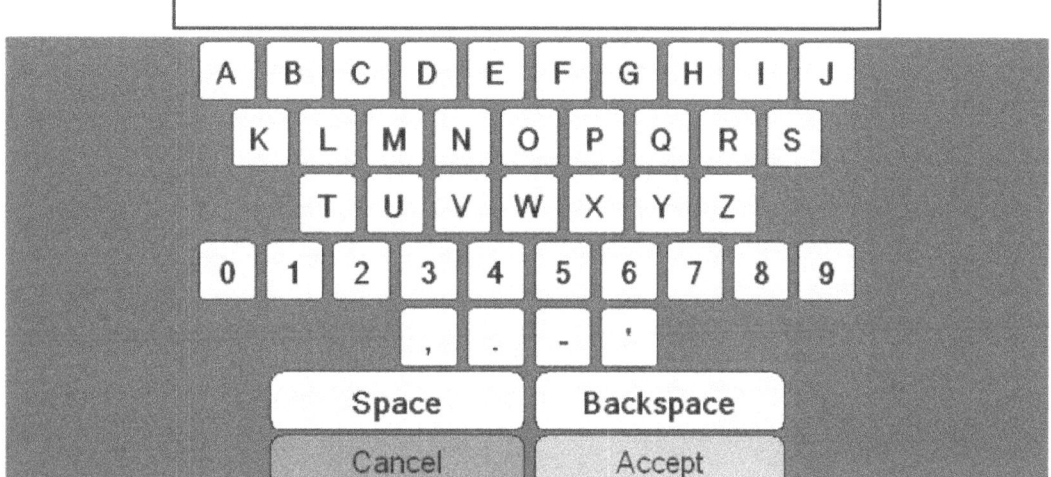

State Senator

Vote for one.

| Edward Shiplett |
| Orange |

| Mary Talarico |
| Tan |

| Write-in Candidate |

Back   (?)   Next

State Assemblyman

Vote for one.

Andrea Solis
Orange

Amos Kellar
Tan

Write-in Candidate

| Back | ? | Next |

County Commissioners

Vote for no more than five.

Camille Argent Orange
Chloe Witherspoon Orange
Clayton Bainbridge Orange
Valerie Altman Orange
Damian Rangel Yellow
Mary Tawa Yellow
Sheila Moskowitz Yellow
Amanda Marracini Tan
Charlene Hennessey Tan
Eric Savoy Tan
Write-in Candidate
Write-in Candidate
Write-in Candidate
Write-in Candidate
Write-in Candidate

Back   ?   Next

City Council

Vote for no more than four.

| Carol Shry |
| Orange |

| Harvey Eagle |
| Orange |

| Randall Rupp |
| Orange |

| Barbara Barker |
| Tan |

| Donald Davis |
| Tan |

| Hugh Smith |
| Tan |

| Reid Feister |
| Tan |

Write-in Candidate

Write-in Candidate

Write-in Candidate

Write-in Candidate

Back (?) Next

Non-partisan offices

Back ? Next

Water Commissioners

Vote for no more than two.

Gregory Seldon

Orville White

Write-in Candidate

Write-in Candidate

Back (?) Next

Court of Appeals Judge

Vote for one.

Michael Marchesani

Write-in Candidate

Back   (?)   Next

Retention Question

State Supreme Court Chief Justice. Shall
Robert Demergue be retained?

Yes

No

Press Yes if you want the official to be
retained in office.

Press No if you do not want the official to
be retained in office.

| Back | (?) | Next |

Retention Question

State Supreme Court Associate Justice.
Shall Margaret Sharp be retained?

Yes

No

Press Yes if you want the official to be
retained in office.

Press No if you do not want the official to
be retained in office.

| Back | ? | Next |

Constitutional Amendment H

Shall there be an amendment to the State constitution allowing the State legislature to enact laws limiting the amount of damages for noneconomic loss that could be awarded for injury or death caused by a health care provider? "Noneconomic loss" generally includes, but is not limited to, losses such as pain and suffering, inconvenience, mental anguish, loss of capacity for enjoyment of life, loss of consortium, and other losses the claimant is entitled to recover as damages under general law.

This amendment will not in any way affect the recovery of damages for economic loss under State law. "Economic loss" generally includes, but is not limited to, monetary losses such as past and future medical expenses, loss of past and future earnings, loss of use of property, costs of repair or replacement, the economic value of domestic services, loss of employment or business opportunities. This amendment will not in any way affect the recovery of any additional damages known under State law as exemplary or punitive damages, which are damages allowed by law to punish a defendant and to deter persons from engaging in similar conduct in the future.

Accept

Reject

Back (?) Next

Ballot Measure 101

Requires primary elections where voters may vote for any state or federal candidate regardless of party registration of voter or candidate. The two primary-election candidates receiving most votes for an office, whether they are candidates with no party or members of same or different party, would be listed on general election ballot. Exempts presidential nominations. Fiscal Impact: No significant net fiscal effect on state and local governments.

Accept

Reject

Back   ?   Next

Ballot Measure 106

Allows individual or class action "unfair business" lawsuits only if actual loss suffered; only government officials may enforce these laws on public's behalf. Fiscal Impact: Unknown state fiscal impact depending on whether the measure increases or decreases court workload and the extent to which diverted funds are replaced. Unknown potential costs to local governments, depending on the extent to which diverted funds are replaced.

Accept

Reject

Back (?) Next

Ballot Summary

Press the candidate name or contest title to return to that part of the ballot. Contests that have red messages have not been completely voted. If satisfied with current selections, press "Next." Otherwise, press "Back."

| Press here to cast your ballot now |

President and Vice President
Adam Cramer and Greg Vuocolo

US Senate
Lloyd Garriss

US Representative
Bruce Reeder

Governor
Gerard Harris

Lieutenant Governor
Anthony Parks

Registrar of Deeds
(HERBERT LIDDICOAT)

State Senator
Edward Shiplett

State Assemblyman
Andrea Solis

County Commissioners
Eric Savoy, Charlene Hennessey, Amanda Marracini

City Council
Donald Davis, Barbara Barker, Carol Shry, Hugh Smith

Water Commissioners
Gregory Seldon, Orville White

Court of Appeals Judge
Michael Marchesani

Retention -- Chief Justice
Yes

Retention -- Associate Justice
Yes

Constitutional Amendment H
Accept

Ballot Measure 101
Reject

Ballot Measure 106
Accept

| Back | ? | Next |

Confirm

The voter may now press "Confirm" to finish casting the ballot.

The voter may press **Return to ballot** if the voter desires to make any more changes or selections.

Return to ballot	Confirm

Thank You

Appendix 3

Ballot B

Official Ballot

Official Ballot for the general election
City of Shoreland
Franklin County, West Virginia

November 4, 2008

Congressional District A
Senate District B
Assembly District LL
Council Districts 2, 4, 6

Next

How to Vote:

To vote for the candidate of your choice, touch that person's name. It will turn yellow.

To write in a candidate: To vote for a person who is not on the ballot, touch **Write in a candidate's name**. You will get more instructions on how to complete your write-in.

If you make a mistake or want to change a vote, first touch the yellow box you no longer want. That box will turn gray. Then, touch the choice you do want.

Back   Next

Straight Party Voting

You can vote all at once for all the candidates of one political party for all the races where the candidates belong to a specific party. (This is called a straight-party ticket.)

If you want most candidates from one party but some candidates from another party, you can vote straight party here and change your vote later at a specific race.

To vote straight party, touch the party name and touch [Next].

To not vote straight party, just touch [Next.]

| Aqua |
| Lime |
| Purple |
| Silver |

[Back] (?) [Next]

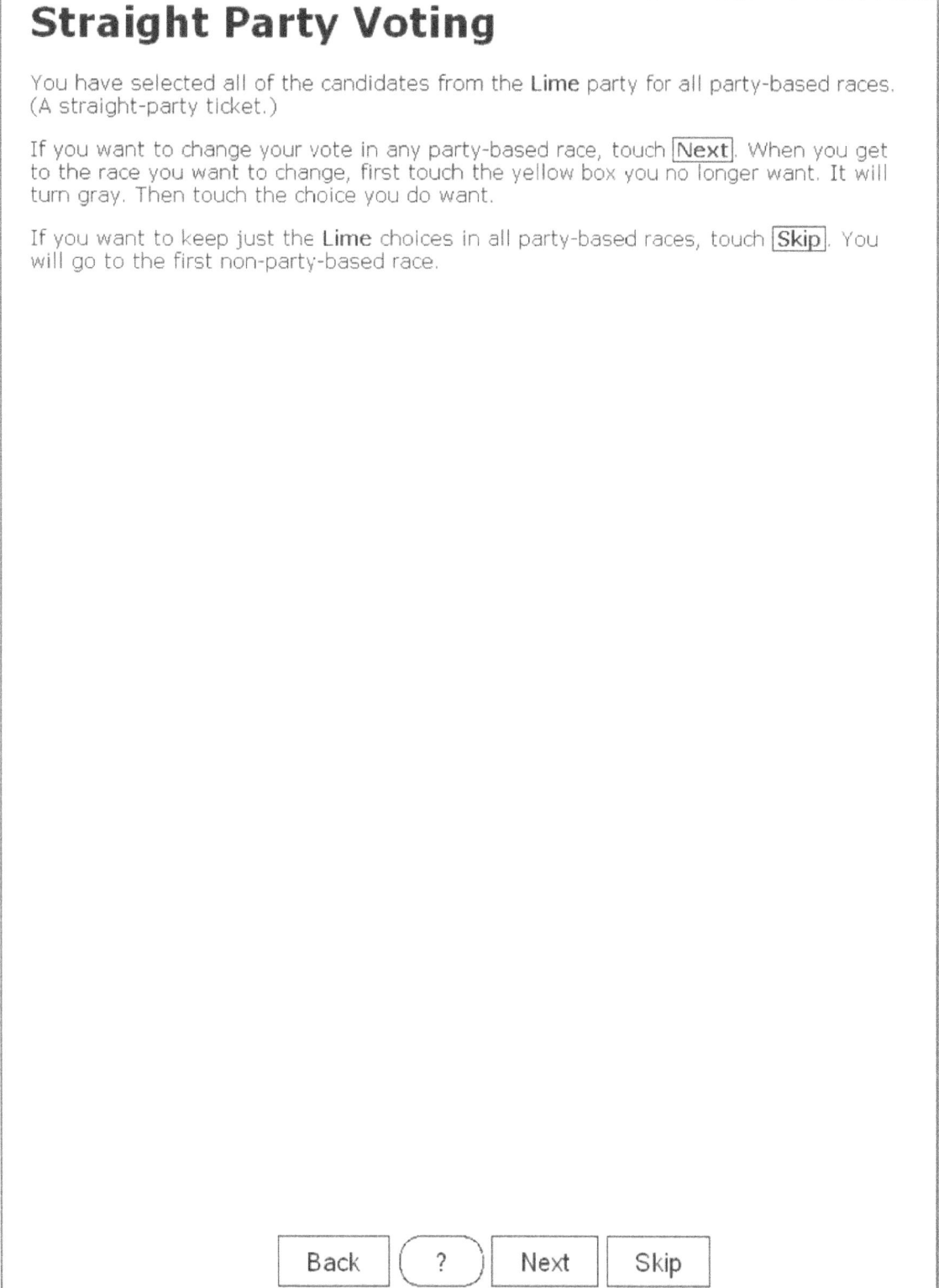

Straight Party Voting

You have selected all of the candidates from the **Lime** party for all party-based races. (A straight-party ticket.)

If you want to change your vote in any party-based race, touch Next . When you get to the race you want to change, first touch the yellow box you no longer want. It will turn gray. Then touch the choice you do want.

If you want to keep just the **Lime** choices in all party-based races, touch Skip . You will go to the first non-party-based race.

Back   (?)   Next   Skip

President and Vice President

Vote for one.

Martin Patterson and Clay Lariviere	Purple
Elizabeth Harp and Antoine Jefferson	Lime
Charles Layne and Andrew Kowalski	Aqua
Marzena Pazgier and Welton Phelps	Silver

Back   (?)   Next

US Senate

Vote for one.

| Victor Martinez |
| Purple |

| David Platt |
| Lime |

| Heather Portier |
| Aqua |

| Write in a candidate's name |

Back (?) Next

US Representative

Vote for one.

| Glen Tawney |
| Purple |

| Carrol Forrest |
| Lime |

Write in a candidate's name

Back (?) Next

Governor

Vote for one.

Fredrick Sharp	Purple

Alex Wallace	Lime

Barbara Williams	Aqua

Althea Sharp	Silver

Ann Windbeck	Independent

Write in a candidate's name

Back	(?)	Next

Lieutenant Governor

Vote for one.

Charles Qualey	Purple
George Hovis	Lime
Burt Zirkle	Aqua
Write in a candidate's name	

Back (?) Next

Registrar of Deeds

Vote for one.

Lillian Cohen
Lime

Write in a candidate's name

Back　(?)　Next

Write In a Candidate

Use this screen to vote for a person who is not on the ballot.

Do not write in someone whose name is already on the ballot for this race.

To write in a candidate:

- Type the person's first and last names.
- Put a blank space between the first and last name by touching Space.
- To erase, touch Backspace.

To complete the write-in, touch Accept.

If you change your mind, touch Cancel.

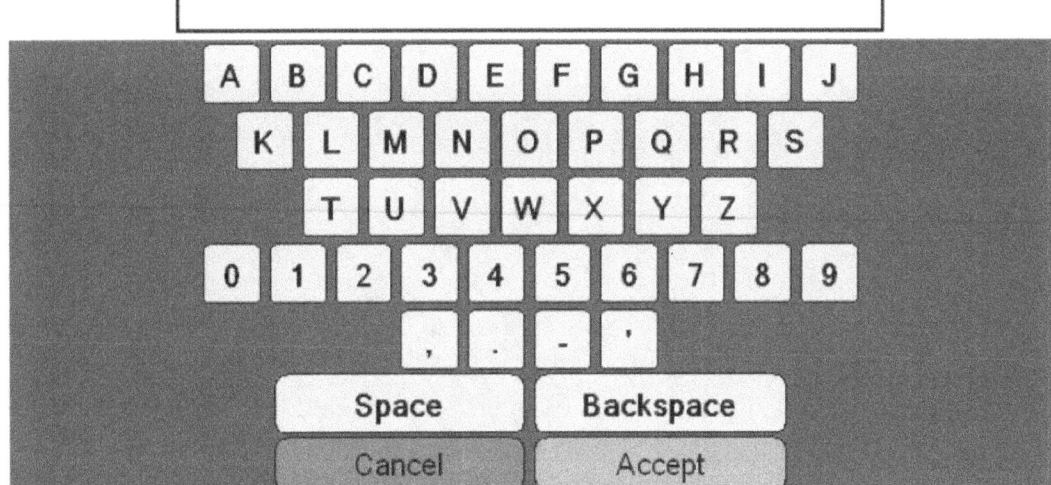

State Senator

Vote for one.

Andrea Solis
Purple

Amos Keller
Lime

Write in a candidate's name

Back (?) Next

State Assembly

Vote for one.

| Edward Shipett |
| Purple |

| Marty Talarico |
| Lime |

Write in a candidate's name

Back   (?)   Next

County Commissioners

Vote for one, two, three, four, or five.

Damian Rangel	
Purple	
Mary Tawa	
Purple	
Sheila Moskowitz	
Purple	
Helen Moore	
Lime	
John White	
Lime	
Valeri Altman	
Lime	
Joe Barry	
Aqua	
Joe Lee	
Aqua	
Martin Schreiner	
Aqua	
Eric Savoy	
Silver	

Write in a candidate's name

Write in a candidate's name

Write in a candidate's name

Write in a candidate's name

Write in a candidate's name

Back   ?   Next

City Council

Vote for one, two, three, or four.

Carole Smith Purple
Randall Eagle Purple
Reid Davis Purple
Barbara Shry Lime
Donald Rupp Lime
Harvey Barker Lime
Hugh Feister Lime
Write in a candidate's name
Write in a candidate's name
Write in a candidate's name
Write in a candidate's name

Back   (?)   Next

Non-partisan offices

If you voted a straight-party ticket, you have not voted for any race from here to the end of the ballot.

Back   (?)   Next

Water Commissioners

Vote for one or two.

Gregory Seldon
Orville White
Write in a candidate's name
Write in a candidate's name

Back   (?)   Next

Court of Appeals Judge

Vote for one.

Kenneth Mitchell

Write in a candidate's name

Back (?) Next

State Supreme Court Chief Justice

Keep **Esther York** as State Supreme Court Chief Justice?

To keep Esther York, touch Yes.

To not keep Esther York, touch No.

Yes

No

| Back | ? | Next |

State Supreme Court Associate Justice

Keep **Elmer Hull** as State Supreme Court Associate Justice?

Yes

To keep Elmer Hull, touch Yes.

No

To not keep Elmer Hull, touch No.

Back   (?)   Next

Constitutional Amendment K

Shall there be an amendment to the State constitution authorizing Madison and Fromwit Counties to hold referenda on whether to authorize slot machines in existing, licensed parimutuel facilities (thoroughbred and harness racing, greyhound racing, and jai alai) that have conducted during each of the last two calendar years before effective date of this amendment? The Legislature may tax slot machine revenues, and any such taxes must supplement public education funding statewide. Requires implementing legislation.

This amendment alone has no fiscal impact on government. If slot machines are authorized in Madison and Fromwit counties, governmental costs associated with additional gambling will increase by an unknown amount and local sales tax-related revenues will be reduced by $5 million to $8 million annually. If the Legislature also chooses to tax slot machine revenues, state tax revenues from Madison and Fromwit counties combined would range from $200 million to $500 million annually.

For

Against

| Back | ? | Next |

Ballot Measure 101

Allows individual or class action "unfair business" lawsuits only if actual loss suffered; only government officials may enforce these laws on public's behalf. Fiscal Impact: Unknown state fiscal impact depending on whether the measure increases or decreases court workload and the extent to which diverted funds are replaced. Unknown potential costs to local governments, depending on the extent to which diverted funds are replaced.

For

Against

| Back | ? | Next |

Ballot Measure 106

Requires primary elections where voters may vote for any state or federal candidate regardless of party registration of voter or candidate. The two primary-election candidates receiving most votes for an office, whether they are candidates with no party or members of same or different party, would be listed on general election ballot. Exempts presidential nominations. Fiscal Impact: No significant net fiscal effect on state and local governments.

For

Against

Back   ?   Review

Review Your Choices

This screen shows everything you voted for. Review it carefully.

Pay attention to any red area. Red means you did not vote or did not vote for as many candidates as you could.

To make changes
1. Touch the race you want to change.
2. At that race, if you have selected something before, touch the choice you do not want. That box will turn gray.
3. Then touch the choice you want.
4. To return to this screen, touch Return to Review .

If you are ready to cast your ballot, touch Ready to Cast Ballot .

Ready to Cast Ballot

President and Vice President
Elizabeth Harp and Antoine Jefferson

US Senate
David Platt

US Representative
Carrol Forrest

Governor
Alex Wallace

Lieutenant Governor
George Hovis

Registrar of Deeds
(ALBERT STERNER)

State Senator
Andrea Solis

State Assembly
Edward Shipett

County Commissioners
Valeri Altman, John White, Helen Moore

City Council
Donald Rupp, Barbara Shry, Carole Smith, Harvey Barker

Water Commissioners
Gregory Seldon, Orville White

Court of Appeals Judge
Kenneth Mitchell

Retention -- Chief Justice
Yes

Retention -- Associate Justice
Yes

Constitutional Amendment K
For

Ballot Measure 101
Against

Ballot Measure 106
For

Back   ?   Next

Confirm

Are you sure you have finished voting?

Note: Once you touch $\boxed{\text{Cast Vote}}$, you will not be able to make any more changes.

If you want to make changes, touch $\boxed{\text{Return to Ballot}}$.

If you are ready to cast your ballot, touch $\boxed{\text{Cast Vote}}$.

$\boxed{\text{Return to Ballot}}$ $\boxed{\text{Cast Vote}}$

Thank You

Your vote has been recorded.
Thank you for voting.

Appendix 4

Screening questionnaire (script for the screener)

Recruiter: We want our participant group to have a mix of characteristics, with most having high school education (and not higher).

Recruiter Script:

Hello, my name is _____, calling for Redish & Associates.

We are recruiting US citizens to participate in a usability study of voting instructions. The results of this study will be used to help design ballot and voting machine instructions that are easy to understand.

We would like to ask you a few questions to see if you are a candidate for this study and if you would like to participate. This will only take a few minutes of your time and no one will attempt to sell you anything. This is strictly for research purposes. If you are interested and meet the study criteria, you will be paid to participate. May I ask you a few questions?

1. Are you a US citizen?
   ☐ Yes
   ☐ No (Exclude)

2. Age
   ☐ 18-21 (if younger than age 18, exclude)
   ☐ 22-25
   ☐ 26-30
   ☐ 31-40
   ☐ 41-50
   ☐ 51-60
   ☐ Over 60

   [Recruiter: We're looking for a mix of ages. Select 2-3 in at least 6 categories]

3. What is the highest level of education that you have completed?

☐ Less than a high school degree   (select 4-5)

☐ High school   (select 4-5)

☐ Some college   (select 2-3)

☐ College degree   (select 2-3)

☐ Some advanced courses beyond college   (select 1-2)

☐ Advanced degree   (Exclude)

☐ Other, please describe _____ (hold for consideration)

[Recruiter: Eliminate people who have advanced degrees; otherwise, select mostly participants with high school diplomas or less, with some participants having some college and a few with college degrees.]

4. Do you or any of your immediate family work in any of the following situations:

☐ As a poll worker on Election Day or as a worker in another part of the voting process (Exclude)

☐ In voting machine manufacturing (Exclude)

☐ In voting machine development, marketing, or sales (Exclude)

☐ In any other position that is part of the voting process. Please describe _____ (Exclude)

☐ None of the above.

NOTE: This survey contains collection of information requirements subject to the Paperwork Reduction Act. Notwithstanding any other provisions of the law, no person is required to respond to, nor shall any person be subject to penalty for failure to comply with, a collection of information subject to the requirements of the Paperwork Reduction Act. The estimate response time for this survey is three minutes. The response time includes the time for reviewing instructions, searching existing data sources, gathering and maintaining the data needed, and completing and reviewing the collection of information. Send Comments regarding this estimate or any other aspects of this collection of information, including suggestions for reducing the length of this questionnaire, to the National Institute of Standards and Technology, Attn., Sharon Laskowski, by email to Sharon.Laskowski@nist.gov, or by phone on 301-975-4535. The OMB Control No. is 0693-0-0043, and it expires on 7/31/09.

OMB Control No. 0693-0043   Expiration 07/31/09

Appendix 5

Script for moderator

Introduction

I hope you don't mind, but I am going to be using a script in our session today. I am doing this so that I make sure that what I say to each person who is participating in our study is the same.

Thank you for agreeing to participate in this study. Your taking part in the study helps us evaluate how well these ballots work for voters. Your participation in the study is confidential. The data you provide will *not* have your name on it.

Do you have any questions at this point?

Before we get started, we have some paperwork to do. This is an Informed Consent Form.

[Hand the participant the Informed Consent Form.]

Please read it. It explains
what we are studying
what you will do
how the information will be treated
and so on

When you have finished reading it, if you are comfortable with what it says, please sign it.

[Wait while the participant reads and signs.]

I'll sign here now. My signature just says that I saw you read and sign the form.

Ballot 1

Let's move over here now.

[Settle the participant in front of the first ballot.]

Now we're ready to start on the first ballot. Although this situation is similar to voting in a real election, I'm going to ask you to do specific things as you use the ballot.

Here's the list of what I want you to do on this ballot. Please read through it and tell me when you are done reading.

[Wait while the participant reads the list of directions.]

From this point on, treat me like a poll worker. If you have questions or problems with the ballot, ask me, as a poll worker. I'll note your question and if it is appropriate, I will help you.

Dana and I will both be taking notes.

Please talk out loud and tell us everything you are doing and thinking as you go through the ballot. That way, we can get a better idea of how you are going about voting and *why* the ballot works or doesn't work for you.

Are you ready?

Go ahead.

[Participant votes to Summary screen. Use the following two probes when appropriate:

Probe 1: Situation: Participant verbalizes confusion, lack of understanding, or that there is a problem but does not spontaneously elaborate or explain. Probe: Ask for an explanation by reflecting back the participant's words. Examples:]
You just said that something is confusing. What is confusing? or
You just said that you don't understand. What do you not understand?

[Probe 2: Situation: Participant is silent for more than about 30 seconds. (Do not intervene here if the participant is obviously reading. We do not want to interrupt reading.) Probe:]
What are you thinking now?

[At the Summary screen, allow participant to do own reading, review, whatever they want on the screen. Intervene just before participant presses Cast Vote or Next. Give participant next set of directions.]

Before you do that, I have another instruction for you.

[Participant follows that instruction – to vote for Water Commissioners. Participant gets back to Summary screen.

Again allow participant to review for whatever time or action participant wants. Intervene just before participant presses Cast Vote or Next. Give participant the final set of directions.]

And I have one last set of instructions for you.

[Participant follows those directions and ends ballot.]

Thank you. Now, please fill out this form.

[Hand the participant the after-each-ballot questionnaire.]

You just voted Ballot A / B [Say appropriate one for the ballot the participant has just voted.] **Please put an A / B** [Say appropriate one for the ballot the participant has just voted.] in the box on each row that represents your answer.

Ballot 2

Thank you. Let's move over here now.

[Settle the participant in front of the second ballot.]

Imagine now that it is a different year and you have come to vote again. Just as we did for the first ballot, I'm going to give you a list of what I want you to do.

Here is the list of what I want you to do on this ballot. Please read through it and tell me when you are done reading.

[Wait while participant reads list of directions.]

From this point on, once again, treat me like a poll worker. If you have questions or problems with the ballot, ask me, as a poll worker. I'll note your question and if it is appropriate, I will help you.

Dana and I will both be taking notes.

Please talk out loud and tell us everything you are doing and thinking as you go through the ballot. That way, we can get a better idea of how you are going about voting and *why* the ballot works or doesn't work for you.

Are you ready?

Go ahead.

[Participant votes to Summary screen. Use the following two probes when appropriate:

Probe 1: Situation: Participant verbalizes confusion, lack of understanding, or that there is a problem but does not spontaneously elaborate or explain. Probe: Ask for an explanation by reflecting back the participant's words. Examples:]

You just said that something is confusing. What is confusing? or
You just said that you don't understand. What do you not understand?

[Probe 2: Situation: Participant is silent for more than ..... . (Do not intervene here if the participant is obviously reading. We do not want to interrupt reading.) Probe:]

What are you thinking now?

[At the Summary screen, allow the participant to do own reading, review, whatever they want on the screen. Intervene just before participant presses Cast Vote or Next. Give participant next set of directions.]

Before you do that, I have another instruction for you.

[Participant follows that instruction – to vote for Water Commissioners. Participant gets back to Summary screen.

Again allow participant to review for whatever time or action participant wants. Intervene just before participant presses Cast Vote or Next. Give participant the final set of directions.]

And I have one last set of instructions for you.

[Participant follows those instructions and ends ballot.]

Thank you. Now, please fill out this form.

[Hand the participant the after-each-ballot questionnaire again.]

You just voted Ballot A / B [Say appropriate one for the ballot the participant has just voted.] Please put an A / B [Say appropriate one for the ballot the participant has just voted.] in the box on each row that represents your answer.

Also, if you want to change your answer for the first ballot that you did, you may do so. Cross out the one you don't want and put the letter A / B [Say appropriate one for the first ballot the participant voted.] in the box you now want for that answer.

Comparative interview with forced choice

Thank you. Let's move over here now. [Settle the participant in the interview space.]

Thank you very much for voting both of those ballots.

I would like to go over them in more detail with you now.

I am going to show you some of the pages from the two ballots. I will show you the same page from both ballots at one time. On each set of pages, I want you to compare the instructions and comment on them.

[Have a stack of each ballot, with the one on the left being the one that the participant voted first. Start by turning over the page of instructions for both ballots and pointing out which is Ballot A and which is Ballot B.]

Notice that the instructions on these pages are different. Please compare them and comment on them.

[Wait for participant's comments. When participant stops:]

Thank you for your comments. Do you have anything else you would like to say about these two pages?

[Wait for participant's comments. If participant makes more comments, repeat cycle of "anything else" prompt. When participant says "no":]

If you had to choose one of these two pages for a ballot, which would you choose?

[Repeat for all pages with differences in instructions. These 16 pages are:
Instructions to Voters / How to Vote
Straight Party Vote / Straight Party Voting (page with parties listed)
Straight Party Vote / Straight Party Voting (page with message after voting)
President and Vice President
Registrar of Deeds (Write-in Candidate / Write in a candidate's name)
Write-In Instructions / Write In a Candidate
County Commissioners
City Council
Non-partisan offices
Water Commissioners
Retention Question / State Supreme Court Justice
Amendment H / Amendment K (response options differ; "good" idea)
Measure 101 (response options differ; "bad" idea)
Ballot Summary / Review Your Choices
Confirm
Thank You]

Closing and demographic / voting experience questionnaire

That was very helpful. Thank you. We are just about finished.

First, I have a small form I would like you to fill out.

[Give participant the one question end-of-session (preference) form.]

We would also like to know a little more about you and about your voting experience. Please answer the questions in this questionnaire as truthfully as you can.

[Give participant the demographic and voting experience questionnaire.]

Give incentive and get signature

Finally, we are happy to give you this envelope with the payment.

[Give participant envelop and form.]

I must ask you to sign this form, acknowledging that you received the payment because I must account for the money we are giving out today.

Thank you and exit

Thank you again. We really appreciate your coming to help with this study.

Appendix 6

Informed Consent Form

What we are studying

As part of the Help American Vote Act (HAVA), the National Institute of Standards and Technology (NIST) is developing guidelines for how to best give instructions to voters on ballots.

We are conducting this study to develop those guidelines. We will use the results of this study to create guidelines and examples to help election officials write useful instructions on ballots.

The Election Assistance Commission (EAC) and NIST are funding this study. Redish & Associates, Inc. and NIST are conducting the study.

What you will be asked to do

You will be with us for approximately one hour.

We have set up ballots with typical voting situations, but the names of the candidates and the parties are made up. For some parts of the ballot, we will tell you whom to vote for or what to do. For other parts of the ballot, you will decide and tell us what you are going to do. You will vote on two different ballots, first one and then the other. While you are voting, we will ask you to say out loud whatever you are reading, doing, or thinking.

After you have voted on the two ballots, we will talk with you about the ballots. And we will ask your opinion about the ballots that you used.

At the end, we will ask you for information about yourself, such as your age, gender, education level, and your experience with voting and voting equipment. This information helps us understand how the results of our study relate to different age groups, and so on.

In addition to collecting your votes, your comments on the ballots, and your information, we are recording the ballot screens and your voice. We are not photographing or recording your face.

How your information will be treated

The data will be collected anonymously and not linked to your name. All data will be identified and linked together by a number and not by your name. We will not have or use your name in any of the data or the reporting.

Your identity will be protected to the extent permitted by law, including the Freedom of Information Act. The people who will be able to review individual records from the study will be limited to

- members of the NIST Institutional Review Board (IRB)
- appropriate project researchers, including NIST employees and Redish & Associates, Inc. employees
- appropriate NIST researchers
- other appropriate Federal employees.

How many people will be in the study

We expect about 50 people to take part in this study.

You are participating voluntarily

You may leave the study at any time.

What risks and benefits there are for you in this study

There are no risks to you in participating in this study.

Nor are there any immediate benefits.

By participating in this study, you are helping us develop long-term benefits for all voters. The long-term benefits of this study should be better voting systems.

Compensation

You will be paid $75 in cash for your participation in this study before you leave.

For more information

For questions about this study, please contact:

Dr. Sharon Laskowski

301 975 4535

sharon.laskowski@nist.gov

For questions about your rights as a person participating in a study (a human subject), please contact:

Lisa R. Karam

Acting NIST IRB Chairperson

301 975 5561 or 301 975 3190

lisa.karam@nist.gov

Your agreement to participate

Please read the following and sign if you are willing to participate in this study.

"I have read the above description of this research project.

I have also spoken to the project researcher, who answered any questions I had about this project.

I agree to participate in this research and I understand that I may withdraw at any time."

Signature: _____ Date: _____

Project researcher name: _____Ginny Redish_____

Project researcher signature: _____ Date: _____

Appendix 7

Directions to participants for voting

Note: When the participant sat down in front of the tablet PC for Ballot A, the moderator gave the participant the sheet of directions for Ballot A and instructed the participant to read through the instructions before beginning to vote and then to keep the instructions to refer to while voting.

Directions for Ballot A

You usually vote for everyone in the Tan party. Vote for all the people in that party at one time.

Even though you voted for everyone in the Tan party, for Registrar of Deeds, you want Herbert Liddicoat. Vote for him.

For State Senator, instead of the Tan party person, you want the Orange party person. Make sure your vote for State Senator is for the Orange party person.

For City Council, you think that the women running are the best candidates, so vote for them. You decide what to do about the other candidates for City Council and tell me what you are doing when you decide.

For now, you decide not to vote for Water Commissioners.

For Court of Appeals Judge, vote for Michael Marchesani.

You don't have a strong feeling about the state Supreme Court justices, so you decide to allow them to stay in office.

You think Constitutional Amendment H is a good idea.

You think Ballot Measure 101 is a bad idea.

You think Ballot Measure 106 is a good idea.

-- end of page of directions that the participant saw before voting –

Note: When the participant was on the Ballot Summary page and about to cast the vote, the moderator intervened, handing the participant the following direction on a separate piece of paper:

You decide that you should vote for the Water Commissioners, so do that now.

Note: When the participant completed the previous direction and was again ready to cast the vote, the moderator intervened for a final time, showing the participant the following directions on a piece of paper:

You realize that you actually wanted Andrea Solis to be your State Assembly person. Change your vote for State Assembly to Andrea Solis.

When you are ready, finish voting as you really would in a real election.

Note: When the participant sat down in front of the tablet PC for Ballot B, the moderator gave the participant the sheet of directions for Ballot B and instructed the participant to read through the instructions before beginning to vote and then to keep the instructions to refer to while voting.

Directions for Ballot B

You usually vote for everyone in the Lime party. Vote for all the people in that party at one time.

Even though you voted for everyone in the Lime party, for Registrar of Deeds, you want Albert Sterner. Vote for him.

For State Senator, instead of the Lime party person, you want the Purple party person. Make sure your vote for State Senator is for the Purple party person.

For City Council, you think that the women running are the best candidates, so vote for them. You decide what to do about the other candidates for City Council and tell me what you are doing when you decide.

For now, you decide not to vote for Water Commissioners.

For Court of Appeals Judge, vote for Kenneth Mitchell.

You don't have a strong feeling about the state Supreme Court justices, so you decide to allow them to stay in office.

You think Constitutional Amendment K is a good idea.

You think Ballot Measure 101 is a bad idea.

You think Ballot Measure 106 is a good idea.

-- end of page of directions that the participant saw before voting –

Note: When the participant was on the Review Your Choices page and about to cast the vote, the moderator intervened, handing the participant the following direction on a separate piece of paper:

You decide that you should vote for the Water Commissioners, so do that now.

Note: When the participant completed the previous direction and was again ready to cast the vote, the moderator intervened for a final time, showing the participant the following directions on a piece of paper:

You realize that you actually wanted Edward Shipplett to be your State Assembly person. Change your vote for State Assembly to Edward Shipplett.

When you are ready, finish voting as you really would in a real election.

Appendix 8

Questions after voting each ballot

P # _____

Please mark how you feel about each statement for the ballot that you just voted. Use the letter (A or B) for the ballot that you just voted.

For each statement, please put the letter in the box on that row that matches how you feel about the statement.

I will give you this form again after you have voted the second ballot.

	Strongly Disagree				Strongly Agree
	1	2	3	4	5
I feel confident that I used this ballot correctly.					
I think that I would need to ask questions to know how to use this ballot.					
I think that most people would figure out how to use this ballot very quickly.					
Figuring out how to vote with this ballot was difficult.					
I think that this ballot was easy to use.					

OMB Control No. 0693-0043   Expiration 07/31/09

NOTE: This survey contains collection of information requirements subject to the Paperwork Reduction Act. Notwithstanding any other provisions of the law, no person is required to respond to, nor shall any person be subject to penalty for failure to comply with, a collection of information subject to the requirements of the Paperwork Reduction Act. The estimate response time for this survey is one minute. The response time includes the time for reviewing instructions, searching existing data sources, gathering and maintaining the data needed, and completing and reviewing the collection of information. Send Comments regarding this estimate or any other aspects of this collection of information, including suggestions for reducing the length of this questionnaire, to the National Institute of Standards and Technology, Attn., Sharon Laskowski, by email to Sharon.laskowski@nist.gov, or by phone on 301-975-4535. The OMB Control No. is 0693-0-0043, and it expires on 7/31/09.

OMB Control No. 0693-0043   Expiration 07/31/09

Appendix 9

End of Session Question

(Overall preference)

(with response numbers added as a right column)

P # _____

The two ballots that you used today had the same races and the same number of candidates and amendments / measures, but they had different placement and wording for the instructions on some of the pages.

If you were in a real voting situation, would you prefer one of these sets of instructions over the other?

		Responses
☐	Yes, I would prefer **Ballot A.**	4
☐	Yes, I would prefer **Ballot B.**	37
☐	No, I have no preference.	4

Please tell us why you chose that answer:

_____

_____

_____

NOTE: This survey contains collection of information requirements subject to the Paperwork Reduction Act. Notwithstanding any other provisions of the law, no person is required to respond to, nor shall any person be subject to penalty for failure to comply with, a collection of information subject to the requirements of the Paperwork Reduction Act. The estimate response time for this survey is one minute. The response time includes the time for reviewing instructions, searching existing data sources, gathering and maintaining the data needed, and completing and reviewing the collection of information. Send Comments regarding this estimate or any other aspects of this collection of information, including suggestions for reducing the length of this questionnaire, to the National Institute of Standards and Technology, Attn., Sharon Laskowski, by email to Sharon.Laskowski@nist.gov, or by phone on 301-975-4535. The OMB Control No. is 0693-0-0043, and it expires on 7/31/09.

OMB Control No. 0693-0043   Expiration 07/31/09

Appendix 10

Guidelines for a Plain Language Ballot

These guidelines are based on the results of an empirical study comparing a ballot with traditional language instructions (Ballot A) to a ballot with plain language instructions (Ballot B).

Voters were more accurate voting the ballot with plain language instructions. Voters preferred the ballot with plain language instructions by a wide margin (82%).

What to say and where to say it

1. Be specific. Give people the information they need.

2. At the beginning of the ballot, explain how to vote, how to change a vote, and that voters may write in a candidate.

3. Put instructions where voters need them. For example, save the instructions on how to use the write-in page for the write-in page.

4. Include information that will prevent voters from making errors, such as a caution to not write in someone who is already on the ballot.

5. On a DRE, never have a page with only a page title (such as the Ballot A page that just said Non-partisan offices).

6. Make the page title the title of the office (State Supreme Court Chief Justice rather than Retention Question).

7. Have voters confirm that they are ready to cast their vote with a Cast Vote button, not a Confirm button.

8. At the end, tell people that their vote has been recorded.

How to say it

9. Write short sentences.

10. Use short, simple, everyday words. For example, do not use "retention" and "retain." Use "keep" instead. For another example, use "for" and "against" for amendments and measures rather than "accept" and "reject."

11. Do not use voting jargon ("partisan" "non-partisan") unless the law requires you to do so. If the law requires these words, work to change the law. Instead refer to contests as "party-based" and "non-party-based."

12.   Address the reader directly with "you" or the imperative ("Do x.").

13.   Write in the active voice, where the person doing the action comes before the verb.

14.   Write in the positive. Tell people what to do rather than what not to do.

15.   Put context before action, "if" before "then." For example, To vote for the candidate of your choice, touch that person's name.

16.   When you want people to act, focus on verbs rather than nouns. For example, Write in a candidate's name.

17.   When giving people instructions that are more than one step, give each step as an item in a numbered list.

18.   Do not number other instructions. When the instructions are not sequential steps, use separate paragraphs with bold beginnings instead of numbering.

19.   Put information in the order that voters need it. Don't tempt voters to irrevocable actions before explaining the other options. (See, for example, the order of the information on the Ballot B Confirm page: a question, a note about consequences, an instruction on how to make changes, and then the irrevocable action described last.)

What to make it look like

20.   Break information into short sections that each cover only one point.

21.   Keep paragraphs short. A one-sentence paragraph is fine.

22.   Separate paragraphs by a space so each paragraph stands out on the page.

23.   Do not use italics.

24.   Use bold for page titles.

25.   Use bold to highlight keywords or sections of the instructions, but don't overdo it.

26.   Keep all the instructions in the left column. Do not put instructions under the choices for a contest.

27.   Do not use all capital letters for emphasis. Use bold. Write all instructions in appropriate upper case and lower case as you would in regular sentences. If the law requires you to use all capital letters, work to change the law.

28.   Use a sans serif font in a readable type size.